Electromagnetic Induction in the Earth and Moon

ADVANCES IN EARTH AND PLANETARY SCIENCES

Advances in Earth and Planetary Sciences 9

Supplement Issue to Journal of Geomagnetism and Geoelectricity

Electromagnetic Induction in the Earth and Moon

Edited by
U. Schmucker

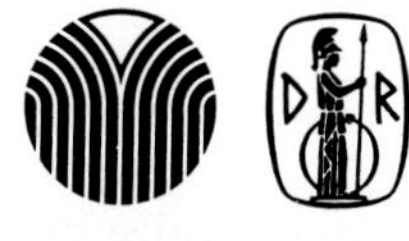

Center for Academic Publications Japan/Tokyo
D. Reidel Publishing Company/Dordrecht · Boston · London

Library of Congress Cataloging in Publication Data

DATA APPEARS ON SEPARATE CARD

ISBN 90-277-1131-3

Published by Center for Academic Publications Japan, Tokyo, in co-publication with D. Reidel Publishing Company, P.O.Box 17, 3300 AA Dordrecht.

Sold and distributed in Japan, China, Korea, Taiwan, Indonesia, Cambodia, Laos, Malaysia, Philippines, Thailand, Vietnam, Burma, Pakistan, India, Bangla Desh, Sri Lanka by Center for Academic Publications Japan, 4-16, Yayoi 2-chome, Bunkyo-ku, Tokyo 113, Japan.

Sold and distributed in the U.S.A. and Canada by Kluwer Boston Inc., Lincoin Building, 160 Old Derby Street, Hingham, MA 02043, U.S.A.

Sold and distributed in all other countries by Kluwer Academic Publishers Group, P.O. Box 322, 3300 AH Dordrecht, Holland.

Printed in Japan

Preface

Physical and chemical studies of the earth and planets along with their surroundings are now developing very rapidly. As these studies are of essentially international character, many international conferences, symposia, seminars and workshops are held every year. To publish proceedings of these meetings is of course important for tracing development of various disciplines of earth and planetary sciences though publishing is getting fast to be an expensive business.

It is my pleasure to learn that the Center for Academic Publications Japan and the Japan Scientific Societies Press have agreed to undertake the publication of a series "Advances in Earth and Planetary Sciences" which should certainly become an important medium for conveying achievements of various meetings to the academic as well as non-academic scientific communities. It is planned to publish the series mostly on the basis of proceedings that appear in the Journal of Geomagnetism and Geoelectricity edited by the Society of Terrestrial Magnetism and Electricity of Japan, the Journal of Physics of the Earth by the Seismological Society of Japan and the Volcanological Society of Japan, and the Geochemical Journal by the Geochemical Society of Japan, although occasional volumes of the series will include independent proceedings.

Selection of meetings, of which the proceedings will be included in the series, will be made by the Editorial Committee for which I have the honour to work as the General Editor. I and the members of the Editorial Committee will certainly welcome any suggestions that will promote the series. Whenever the convener of a meeting related to earth and planetary sciences is in a position to have to look for a medium for publishing the proceedings please contact us.

Tsuneji Rikitake
General Editor

Foreword

Electromagnetic induction by diurnal S_q variations and global stormtime D_{st}-Variations provides a unique opportunity to look deeply into the Earth's mantle, using the electrical conductivity as a diagnostic parameter for the state and composition of mantle material at great depth. On August 24 in 1977 the IAGA Working Group 3/I on EM induction in the Earth and Moon held a full-day symposium on "Geomagnetic Induction by Long-period Variations" during the Joint General Assembly of IAGA and IAMAP at Seattle, Washington/USA. From the twenty papers presented thirteen appear in final form in this supplement issue of the Journal of Geomagnetism and Geoelectricity. It contains in addition three papers which could not be presented because the authors have been unable to come to the conference. One paper has been presented in a different session, but was included in this issue because of its relevance to the topic of the symposium.

The series of papers begins with an invited review by Dr. J. Filloux and is in its first part centered on induction problems connected with the oceans, covering as a well conducting thin sheet of highly irregular shape the less conducting layers of crust and uppermost mantle. Dr. W. D. Parkinson's invited review on the induction by S_q is followed by papers which are likewise concerned with deeply penetrating variations in various parts of the world. They are augmented by papers using the magnetotelluric impedance of faster variations to probe the conductivity distribution of distinct geological structures. I am indebted to Dr. M. Richards for his editorial help.

Ulrich Schmucker
Chairman
IAGA Working Group 3/I

CONTENTS

Preface v
Foreword .... U. SCHMUCKER vii
Observation of Very Low Frequency Electromagnetic Signals in the Ocean .... J. H. FILLOUX 1
Atlantic Lithosphere Sounding .... C. S. COX, J. H. FILLOUX, D. I. GOUGH, J. C. LARSEN, K. A. POEHLS, R. P. VON HERZEN, and R. WINTER 13
North Pacific Magnetotelluric Experiments .... J. H. FILLOUX 33
Electromagnetic Induction in the Vancouver Island Region ....W. NIENABER, H. W. DOSSO, L. K. LAW, F. W. JONES, and V. RAMASWAMY 45
Induction in Arbitrarily Shaped Oceans II: Edge Correction for the Case of Infinite Conductivity ....R. C. HEWSON-BROWNE and P. C. KENDALL 51
The Effect of a Simple Model of the Pacific Ocean on S_q Variations .... B. A. HOBBS and G. J. K. DAWES 59
Electromagnetic Induction at a Model Ocean Coast ....G. FISCHER, P.-A. SCHNEGG, and K. D. USADEL 67
Diakoptic Solution of Induction Problems .... C. R. BREWITT-TAYLOR and P. B. JOHNS 73
Induction by S_q .... W. D. PARKINSON 79
Electromagnetic Response Functions from Interrupted and Noisy Data .... J. C. LARSEN 89
Deep Conductivity Distribution on the Russian Platform from the Results of Combined Magnetotelluric and Global Magnetovariational Data Interpretation .... A. A. KOVTUN and L. N. POROKHOVA 105
Connection between the Electric Conductivity Increase due to Phase Transition and Heat Flow .... A. ÁDÁM 115
Geomagnetic Variations Behavior in Central Europe ....I. I. ROKITYANSKY 125
Geomagnetic Sounding of an Ancient Plate Margin in the Canadian Appalachians .... J. A. WRIGHT and N. A. COCHRANE 133
Magnetovariational and Magnetotelluric Investigations in S. Scotland .... V. R. S. HUTTON and A. G. JONES 141
An Analogue Model Study of Ocean-Wave Induced Magnetic Field Variations Near a Coastline .... T. MILES and H. W. DOSSO 151
Long Period Variations of the Geomagnetic Field and Inferences about the Deep Electric Conductivity of the Earth ....A. M. IŞIKARA 155
Inverse Magnetotelluric Problem for Sounding Curves from Czechoslovak Localities ....K. PĚČ, J. PĚČOVÁ, and O. PRAUS 159

Remarks on Spatial Distribution of Long Period Variations in the Geomagnetic Field over European Area...... J. PĚČOVÁ, K. PĚČ, and O. PRAUS 171
An Interpretation of the Induction Arrows at Indian Stations ....
.... B. J. SRIVASTAVA and H. ABBAS 187

Observation of Very Low Frequency Electromagnetic Signals in the Ocean

J. H. FILLOUX

Scripps Institution of Oceanography, University of California, San Diego, La Jolla, California, U.S.A.

(Received December 12, 1977; Revised February 24, 1978)

Since magnetic signals recordable on the deep sea floor are very small and since records of long duration must be obtained to extend the frequency range abruptly limited on the high end by ocean shielding, sea floor magnetic variographs depend critically on long term stability. Temperature effects are not limiting since temperature variations at great oceanic depth are minimal, but stability of position in situ at landing depends on little known as well as little controllable conditions.

Magnetic variations on the sea floor are presently carried out almost exclusively by means of self contained magnetic variographs (often simply called magnetometers) freely traveling between ocean surface and bottom. The sensors are flux gates, thin-film single domain transducers and suspended magnets with optical readout. In the first two cases, the achieved stability is dependent upon the stability of the cancelling field and associated electronic circuitry. In suspended magnet types, initial torsion of the suspension fibers can relax the requirements upon electronics considerably.

Electric field recording on the sea floor by means of long lines requires a sufficiently large separation to minimize effectively electrode drift. Azimuth and electrode separation must be established with adequate accuracy, a non-trivial requirement. Electrode drift can be eliminated within the frequency range of interest—below a few cycles per minute—by electrode switching. This technique leads to more compact instruments. Its implementation is illustrated in detail.

Meaningful electromagnetic observations related to oceanic signals are also made at the periphery of continents or on islands by means of wires specially laid locally or by means of abandoned transoceanic cables.

1. Introduction

While electromagnetic exploration of the earth's interior remains a complex problem at the data interpretation level, its experimental application has produced a great variety of successful observational methods for land usage. A comprehensive review of the equipment and techniques in use has been presented by SERSON (1973), together with an extended list of references. With respect to the general subject, the significance of global geomagnetic sounding—mainly based on continental data—has been reviewed and referenced by BAILEY (1973), RIKITAKE (1973) and others. Some examples of regional magnetotelluric research on the continent can be found in SCHMUCKER (1973) and GOUGH (1973). Altogether an impressive amount of electromagnetic exploration has been carried out on land.

Electromagnetic exploration of the oceanic basement, in spite of a definite increase in pace, remains limited. The pioneering observational advances in this field accomplished prior to the late sixties were reviewed by FILLOUX (1973) (see also BENNETT and FILLOUX, 1975). Because of staggering initial experimental difficulties, the most conclusive steps are probably the most recent ones. Therefore it appears useful to update at this time our earlier review. This objective is the purpose of the present paper.

In order to set a proper perspective, the following background may be useful. There are considerable differences, in spite of a common principle, between land and sea floor magnetotelluric exploration. In both cases the signals relationships of interest are with respect to electromagnetic sources external to the earth, mainly distributed in the ionosphere and vicinity. However, the signals observed on the sea floor suffer two degrees of degradation: (1) the shielding effect of several kilometers of sea water attenuates drastically the ionospheric signals at frequencies above a few cycles per hour, and (2) oceanic water motions, through their interacting with the main earth's field, create unrelated motional fields which contaminate those induced by ionospheric sources. As a result, ocean floor magnetotelluric observations suffer an unavoidable high frequency cutoff somewhere between 10 and 100 cph; furthermore, they become increasingly noisy at frequencies below .1 cph and beyond, in such a way that very long observations are imperative to obtain data of usable quality at very low frequencies. Fortunately, in spite of these stringent limitations it appears that the restricted effective frequency band of only three decades is nevertheless adequately centered along the frequency scale to permit characterization of one of the most interesting features of submarine geology, namely the transition zone between rigid oceanic plates and the more stationary mantle below.

Other fundamental differences between ocean floor and land magnetotelluric exploration relate to their logistics aspects, to the destructiveness of the ocean and the remoteness of the sea floor. It is fair to say, however, that the quietness of the sea floor, the remarkable stability there of sea water temperature and composition as well as the very low source impedance in electric field monitoring constitute definite long term observational advantages.

A limitation common to land and oceanic magnetotellurics is the problem of electrodes in long term electric field measurements. For oceanic use, silver-silver chloride electrodes are adopted generally. A disquieting situation, however, must be faced. Circumstantial evidence has suggested on occasions that oceanic electric fields measured in the past by means of long lines (FILLOUX, 1967a; COX *et al.*, 1971) may have been underestimated (GREENHOUSE, 1972; LAUNAY, 1974). Recent experiments (DANIEL, 1978) and more specifically our own check on earlier work (FILLOUX, 1980) lead to a similar conjecture. Since we had taken extreme care to insure full stretching of our sea floor cables and had included arrays of compasses to check the correctness of orientation at regular intervals, a cause other than inadequate stretching must be considered. A natural starting point is to question the electrodes and to re-examine their behavior when associated with moderately low impedance circuits as was the case at the time of initial experimentation.

In the following we briefly summarize the significant oceanic electromagnetic exploratory work that, to our knowledge, took place during the present decade.

2. Observations

A great variety of electromagnetic observations were made during the spring of 1973 in the western central north Atlantic on the occasion of the multi-institutional MODE experiment (Mid Ocean Dynamic Experiment; for reference see MODE (1972); location 28°N, 68°W, that is 300 nautical miles to the south-west of Bermuda within a circular area 200 miles in diameter). Although the primary objectives of MODE were aimed at an intensive survey of large scale truly oceanic circulation, electromagnetic experiments related to submarine geology were also performed.

Three types of observations were carried out on the deep sea floor. Namely: (1) magnetic variations by means of single domain thin film magnetometer (POEHLS and VON HERZEN, 1976); (2) horizontal electric field, using for the first time small size free fall recorders with electrode noise and drift rejected by means of salt bridge choppers (FILLOUX, 1974; COX *et al.*, 1976); and (3) vertical electric field (HARVEY, 1972). (Vertical electric field measurements were of interest in MODE because the vertical field is a clean measure of oceanic water motion along the magnetic east-west direction.)

During the same three month period, the three component magnetic variations were recorded by Gough on Bermuda and on Great Abaco, Bahamas, and the electric field was monitored at two stations by Larsen (COX *et al.*, 1980).

Another joint experiment was carried out during 1977 in the North Central Pacific Ocean in an area centered 500 nautical miles to the northeast of Hawaii. In addition to electric and magnetic observations on the sea floor similar to those of MODE and carried out by the same researchers, Daniel used a new approach, engineered by Harvey, to deploy a long and fine wire along the sea floor and Filloux tested newly developed three component magnetic variographs of the suspended magnet type (as described below).

More recently our 1965 station off central California was revisited (FILLOUX, 1967a, b, 1977, 1980) for the purpose of checking our early work with equipment now far more adequate than in the original experiments and also to verify improvements of our salt bridge choppers and suspension magnetometers.

Klein and Larsen pursued their investigations of the deep mid oceanic mantle below the Hawaiian chain, having extended their range of observation from Oahu to Hawaii and Midway (KLEIN, 1975; LARSEN, 1976).

RICHARDS (1977) continues his work related to electromagnetic oceanic signals recorded by means of abandoned transoceanic telegraph cables between several islands of the Pacific Ocean. COX *et al.* (1978) have carried out coastal experiments aimed at the investigation of a postulated wave guide below the ocean assumed to be constituted of a very low conductivity layer sandwiched between highly conducing materials, namely the ocean above and hotter layers below. For related results concerning this research see also PISTEK (1977).

A rather unusual experiment, indirectly related to the subject, was performed by HARVEY *et al.* (1977) in the Straits of Magellan. The purpose of this experiment was a feasibility study for a tidal current monitoring station that would make real time estimates of the actual velocity in the channel, so as to provide help to shipping traffic in this exceedingly dangerous area. The working hypothesis was that the motional field of the tidal currents would drive electric currents closing through the adjacent land thus producing there an electric field sufficiently intense to be observable and interpretable. Simultaneous measurements with current meters located in the channel have substantially confirmed the expectations.

3. Spectral Analysis

In interpreting sea floor magnetotelluric data, and to some extent in planning experiments, we lack a full understanding of the importance of interference by oceanic motional fields. Three types of velocity fields are suspected to contribute to a detrimental noise level: (1) mesoscale motions, with periods from a few days to several weeks and perhaps months; such motions were clearly identified in MODE (COX *et al.*, 1976; FREELAND and GOULD, 1976; MCWILLIAMS, 1976). From measurements elsewhere— California current, north central Pacific—where barotropic mesoscale motions are moderate to low, and eastern north Pacific where they are strong, it is apparent that their associated electric spectra varies greatly from place to place; (2) internal waves, ever present in the open ocean (GARRETT and MUNK, 1975), constitute the most likely cause of degradation of coherence between electric and magnetic fluctuations recorded on the deep sea floor in the range of interest of magnetotelluric exploration; (3) local turbulence of small size has been occasionally detected on the deep sea floor, probably associated with the interaction between bottom flow and bottom roughness (COX *et al.*, 1976).

There are other strong motional signals present on the sea floor; in fact oceanic gravitational tides generate the most conspicuous energy peaks. However, since their driving mechanisms are precisely known, are nearly harmonic and affect only narrow bands their effects are not excessively damaging. Even so, the complication contributed to data analysis by the effect of oceanic tides is far from trivial. Effective reduction of this limitation is best accomplished by means of long term records, as also required to extend the usability of the low frequencies in the presence of active mesoscale motions. The fields associated with local sea floor turbulence can be controlled to some extent if several measurements in the same vicinity, but separated by a distance greater than the detrimental turbulence scale, are carried out simultaneously. This remark is also valid with respect to internal wave contamination; however a greater separation—perhaps 50 to 100 km—is thus necessary, a situation which may not always be compatible with the sea floor lateral variability.

In Figs. 1 and 2 we have updated our earlier spectra of electric and magnetic pulsations on the deep sea floor and at mid latitudes (FILLOUX, 1973). It is likely that these will have to be updated again as we continue our learning. In general, oceanic motions create electromagnetic fields which affect more adversely the electric records than the magnetic. In fact magnetic signals energetic enough to overcome ionospheric fluctuations occur only in the case of tides, in very narrow bands, and possibly also in the case of the most energetic mesoscale motions.

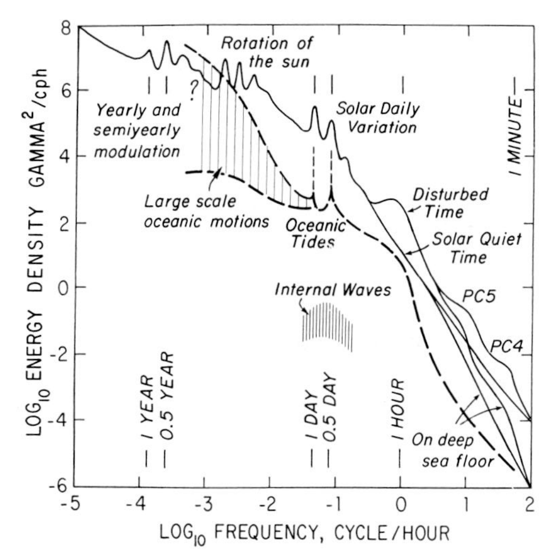

Fig. 1. Spectra of natural magnetic fluctuations observable on earth at mid latitudes. The information related to ionospheric signals distinguishes between solar quiet and magnetically disturbed time. Fluctuation levels are given for the earth's surface and also for the deep sea floor. The frequency band useful to sea floor MT exploration with present instrumentation covers the range .01 to 10 cph. The level of signals related to oceanic motions is also shown.

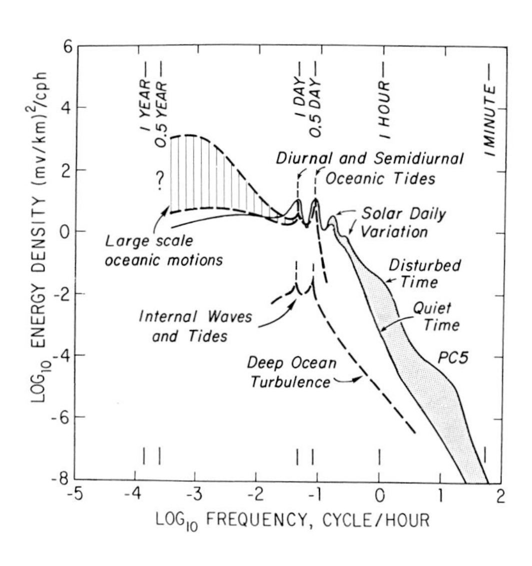

Fig. 2. Spectra of natural electric fluctuations observable on the deep sea floor. Ionospheric signals at mid latitudes for both solar quiet and disturbed time, and motional signals. Notice the importance of the contamination of electric ionospheric signals by motional fields. This effect is much more pronounced here than for the magnetic signals (see Fig. 1).

4. Instrumentation

With respect to instrumentation the important trends and innovations are reviewde in the following. In the practice of sea floor magnetic variation recording we note the abandonment of nuclear resonance devices and of reliance upon thin-film single domain sensors. In spite of the total immunity of the former to temperature effects, the necessity to provide biasing fields to cancel the main earth's field requires high stability electronics and high power, thus tending to renew the problem of temperature compensation. The latter are attractive devices by their sensitivity and simplicity, but they are very temperature dependent—several hundreds of gammas per degree—and their long term stability leaves much to be desired.

Most running projects as well as those in the planning rely on fluxgate sensors. These solid state devices appear to have reached a high degree of quality and produce frequency modulated signals directly adaptable to digital recording techniques. Whether used in nulling configuration or in conjunction with Helmholtz coil cancelling the main earth's field they require a stability of electronics equivalent to the ratio of the desired least count to the main earth's field, that is two parts in a million for a least count of $.1\gamma$. Such a performance in turn requires high class electronic technology not easily amenable to low power consumption.

Our own preference has been for suspension type sensors. We have been guided in this choice by earlier experience with very sensitive and stable optometric angular detectors (FILLOUX, 1967b, 1970) and by our desire to draw on some definite advantages of torsion fibers: in suspended magnet type variographs, cancelation of the main earth's field can be made by twisting the suspension fiber; the required torque orienting the magnets properly is then produced at no drain on the power supply and with no natural noise contribution other than the possible plastic flow or creep in the fiber due to stress; a proper design can be made almost immune to plastic flow by maximizing the angular sensitivity of the optics, optimizing magnet and fiber dimensions and selecting appropriate fiber material.

The general principle of individual suspensions in our three component variographs is shown on Fig. 3. An infrared source generates a light beam collimated by a condens-

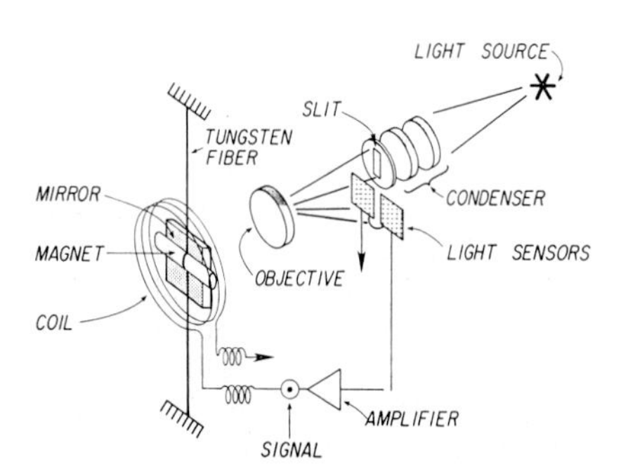

Fig. 3. Principle of our suspension type magnetic variographs. Three such sensors are required for each instrument.

er, truncated by a slit and forming an image onto a mirror attached to the suspended magnet after crossing an objective in front of the mirror. The reflected beam crosses the objective again and forms an image of the slit onto a pair of light sensors connected in opposition. If mirror and magnet are not well centered, a signal results which is amplified and injected into a feedback coil. The magnetic field thus created interacts with the magnet to correctly orient the system. The current in the coil is therefore a measure of the variation field in a direction mutually normal to fiber and magnetic moment of the magnet.

Three identical suspensions are used in each instrument (see illustration of Fig. 4).

Fig. 4. General appearance of our three component sea floor magnetic variographs. The three sensors are mounted in a rotating cage that orients itself a few hours after the bottom has been reached. The tiltmeter-compass is shown at the top. The instrument's heavy aluminum pressure case is seen at right.

They are mounted in a rotating cage that is oriented by a servomechanism when the instrument, dropped from the surface, has reached the bottom. Deviations of up to 15° from ideal orientation are tolerable. The least count has been set to $.2\gamma$, but could be reduced. Power requirement for sensors and recorder is 50 milliwatt. Sampling involving continuous averaging with recording at 128 samples per hour is limited by cassette sizes to 60 days, longer with a reduced sampling rate. The exact orientation of the instrument on the sea floor is provided by the elapsed time shadowgraph of a hanging compass. The compass magnetic needle is enclosed in a transparent capsule with opaque indexing. The capsule hangs as a pendulum from a small ring on the upper lid of a light tight cylindrical box with vertical axis. From the center of the upper lid a point light source is switched on for one second daily. The shadow of compass needle, index and circular capsule is recorded on a photographic film located on the lower lid. This information provides azimuth, tilt and a coarse indication of position stability with time. For illustration of typical recordings see FILLOUX (1980).

While progress has been made in measuring the electric field on the sea floor with instruments of small size, the long line technique has not been abandoned. The technique of Harvey and Daniel illustrated on Fig. 5 and used in the North Pacific experiment of the summer of 1976 (Cruise, Farewell to Aggy) shows that a rather small wire can be laid on the sea floor and can remain insulated for at least several months. The technique involves lowering the recorder by means of the ship's hydrographic wire to within a few hundred meters of the bottom, releasing the instrument and first electrode and letting the wire unspool itself by the motion of the ship towards the position of the second electrode. When the end of the wire is reached, the end electrode falls of its own accord, completing the operation. The distance between deployment rig and bottom is estimated by comparing the dual pluses of a pinger as they are received onboard, namely the direct

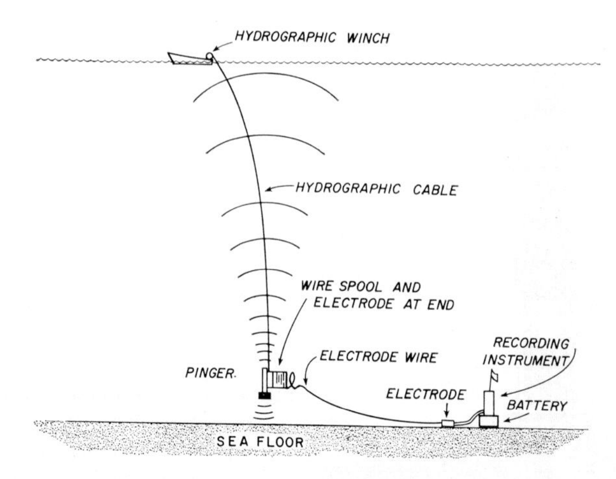

Fig. 5. Technique used by Harvey and Daniel to deploy a long, fine electric conductor along the sea floor.

and the bottom reflected pulses (see Fig. 5). Switching between different pulse rates provides information on the unfolding of the operation. The deployment method has definite merit, but takes time and one operation is required for each component. Since full stretching and intended azimuth cannot be guaranteed, a method permitting precise location of the electrodes must be found. It is our understanding that this requirement shall be fulfilled in future experiments by means of acoustic devices.

Although the art of constructing salt bridge choppers of adequate quality turned out to be difficult to learn, we are confident that this technique has much to offer (for earlier background see MANGELSDORF, 1968; FILLOUX, 1973, 1974). The benefits include small electrode drift and low frequency noise rejection, overall chopper stabilization that eliminates the requirement for regulated supplies, small size and ease of transportation and handling.

The principle of the salt bridge chopper, or electrode switching scheme, is shown on Fig. 6, see legend for detailed operating mechanism. Our initial implementation of

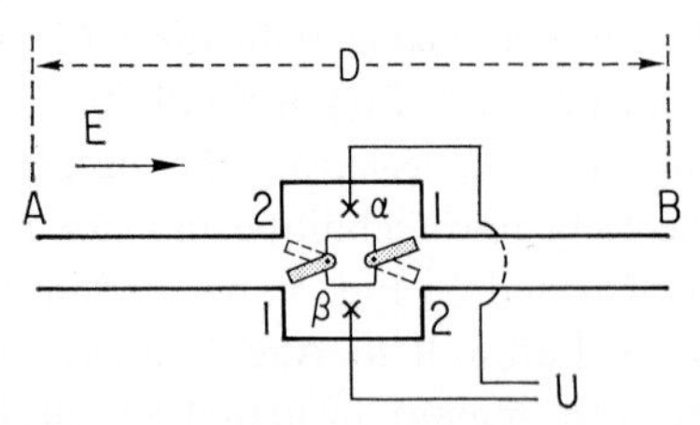

Fig. 6. Principle of the salt bridge chopper. The instrument is schematically represented, submerged in sea water and submitted to electric field E. As shown, the electrode α is connected to the sea at point A, β at point B, and the signal is u1. As the double switch moves from position 1 to 2, the relationship A, B to α, β is inverted and the output becomes u2. The electric field in direction AB is thus $(u2-u1)/2D$ where D is the distance AB and the electrode natural e.m.f., including noise and drift, is $(u2+u1)/2$.

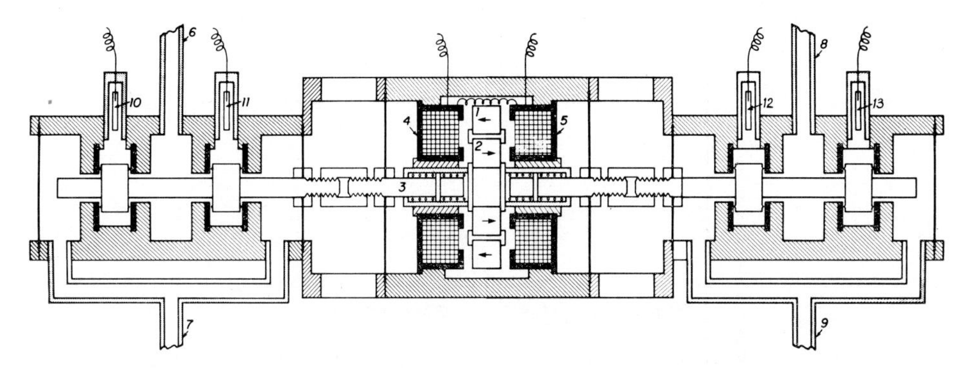

Fig. 7. Functional representation of the twin salt bridge choppers and electromagnetic driver. The system is symmetric with one independent chopper on each side of the central driver. The latter is constituted by two concentric ceramic ring magnets 1 and 2, polarized in a direction parallel to their axis but in opposed directions. They are connected to the sliding shaft, 3, that actuates the choppers, by means of springs (small circles). This compliant connection serves two purposes (1) it relaxes the adverse consequences of differential compression, at abyssal pressure, between metal and plastic parts and (2) it allows storage of otherwise lost magnetic energy during the second half of each switching, to the end of saving electric power; see Fig. 8 and text. The magnet assembly moves within the gap of two electromagnetic armatures 4 and 5. A strong magnetic flux generated by the magnets flows through a low reluctance path constituted by the soft iron armature, thus producing a strong attraction of the magnet assembly, either to the right or to the left, with an unstable neutral position half way in between. The system is immersed in light oil, laterally contained by flexible neoprene membranes. Switching from left to right is produced by a positive current pulse in both coils within the armature. Negative pulses cause switching in the opposite direction in such a way that the chopper position is unambiguously polarity controlled. No current—hence no power—is required to sustain the forces that maintain the water switches active between reversals. The relation between these salt bridges double pull double-throw switches and the switching schemes of Fig. 6 is easily recognizable: the pipes A and B are here shown as 6 and 7 for one chopper and 8 and 9 for the other; the electrodes pairs are represented by 10, 11, and 12, 13.

this system, as used in MODE (Cox *et al.*, 1976), Farewell to Aggy (Filloux, 1977) and S.F. Revisited (Filloux, 1980), is shown on Fig. 7; again the legend provides details on construction and operation for choppers and driver. For response characteristics to both oceanic electric field and electrode noise see Filloux (1974).

The technical problems encountered in developing the salt bridge chopper included the concept of a driving system requiring minimum power, the selection of valve-like switches capable of opening completely (that is, allowing no current leakage) and without degradation with time, and of a design immune to the very large changes in hydrostatic pressure between sea surface and sea floor. The last requirement is of great importance on account of the large differences between elasticity modulii of the insulating materials used in switching components and of the metallic parts of the driver. The difficulties are compounded by the impossibility to visually determine where distortions cause failures as pressure is applied and by the ambiguity encountered in trying to separate aging effects in the switch mechanisms from those due to dimensional changes induced by creep under

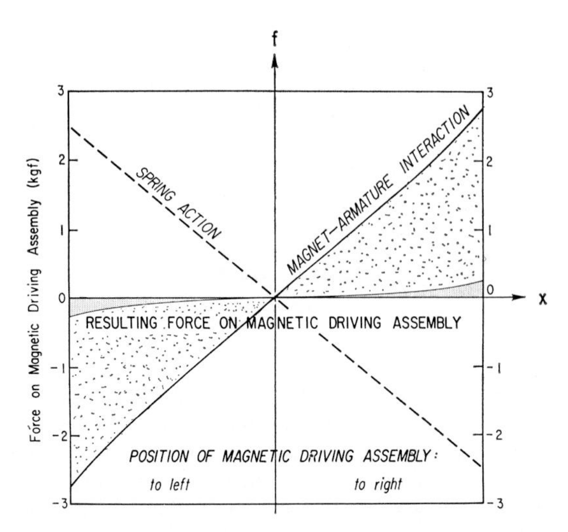

Fig. 8. Combined forces onto magnet assembly of driver as a function of position of magnet assembly within the gap between armatures. Light grey area represents energy required to expel magnet from one side to the other when springs are not used (see Fig. 7); with optimum spring system the energy is reduced to the much smaller dark grey area, with concurrent large energy saving, yet with no decrease in the force on sealing components.

Fig. 9. Launching of a self contained, free-fall sea floor electric field recorder with electrode drift removal by electrode switching. Chopper is seen at left-bottom of vertical cylindrical pressure case. Salt bridge pipes have 6 meter span. Tripod ballast is released for buoyant return. C. B. radio beacons (50 mWatt) facilitate recovery.

abyssal pressure.

The design of the electromechanical driver with its twin ceramic ring magnets and spring system resolves simultaneously several of the aforementioned difficulties, see Fig. 7. In order to maximize the force exerted on the switch seats by the magnet-armature interaction, the gap, or magnet travel, must be large; otherwise the differential force resulting from the magnets interactions with the left and with the right armatures tend to cancel each other. In the absence of the spring system, the switching energy would be proportional to the light grey area in Fig. 8. However, by reducing dramatically the travel in the choppers, but not in the driver, and by selecting springs with a spring constant just slightly less than that compensating the magnetic force on the driving magnets, the driving energy falls considerably and becomes proportional to the dark grey area of Fig. 8. At the same time the initial adjustment of the switch is made much less critical and subsequent changes in dimensions are tolerated unless very large. Typically the measured energy requirement is .25 Joule per switching, .5 Joule per cycle, or 64 Joule/hr at 128 samples per hr. Thus 5 kg of alcaline batteries can provide ample power for one year of operating at 64 samples per hr.

The launching of a two horizontal component sea floor electric field recorder is shown on Fig. 9.

Following a long developmental period, rich in expectations, as well as in hard won technological breakthroughs, a more decisive epoch appears to have been reached in which tooling adequate to electromagnetic exploration of the oceanic lithosphere and upper mantle has become available. We therefore expect increased contributions from MT sea floor exploration to submarine geological research in a not too distant future.

Preparation of this manuscript was supported under NSF Grant OCE76-81867.

REFERENCES

Bailey, R. C., Global magnetic sounding, methods and results, *Phys. Earth Planet. Inter.*, **7**, 234–244, 1973.

Bennett, D. J. and J. H. Filloux, Magnetotelluric deep electrical sounding and resistivity, *Rev. Geophys. Space Phys.*, **13**, 197–203, 1975.

Cox, C. S., J. H. Filloux, and D. Cayan, Electromagnetic observations during MODE–I, Sea floor electric field and transport, *Polymode News*, **14**, 5–10, Aug. 27, 1976.

Cox, C. S., J. H. Filloux, and J. C. Larsen, Electromagnetic studies of ocean currents and electrical conductivity below the ocean floor, in *The Sea*, Vol. 4, Part 1, edited by A. Maxwell, pp. 637–693, Wiley-Interscience, New York, 1971.

Cox, C. S., N. Kroll, P. Pistek, and K. Watson, Electromagnetic fluctuations induced by wind waves on the deep sea floor, *J. Geophys. Res.*, **83**, 431–442, 1978.

Cox, C. S., J. H. Filloux, D. I. Gough, J. C. Larsen, K. A. Poehls, R. P. Von Herzen, and R. Winter, Atlantic lithosphere sounding, *J. Geomag. Geoelectr.*, 1980 (this issue).

Daniel, T., Triaxial *E* field measurements for defermining deep ocean water motions, Ph. D. Thesis, Univ. of Hewaii, 1978.

Filloux, J. H., Oceanic electric currents, geomagnetic variations and the deep electrical conductivity structure of the ocean continent transition of Central California, Ph. D. Thesis, Univ. of Calif., San Diego, 166 pp. 1967a.

Filloux, J. H., An ocean bottom *D* component magnetometer, *Geophysics*, **32**, 978–987, 1967b.

Filloux, J. H., Deep sea tide gauge with optical readout of Bourdon tube rotation, *Nature*, **226**, 925–937, 1970.

Filloux, J. H., Techniques and instrumentation for the study of natural electromagnetic induction at sea, *Phys. Earth Planet. Inter.*, **7**, 323–328, 1973.

Filloux, J. H., Electric recording on the sea floor with short span instruments, *J. Geomag. Geoelectr.*, **26**, 269–279, 1974.

FILLOUX, J. H., Ocean floor magnetotelluric sounding over the North Central Pacific, *Nature*, **269**, 297–301, 1977.

FILLOUX, J. H., North Pacific magnetotelluric experiments: sea floor instrumentation, soundings and discussion, *J. Geomag. Geoelectr.*, 1980 (this issue).

FREELAND, H. J. and W. J. GOULD, Objective analysis of meso-scale ocean circulation features, *Deep-Sea Res.*, **23**, 915–923, 1976.

GARRETT, C. and W. MUNK, Space-time scales of internal waves: A progress report, *J. Geophys. Res.*, **30**, 291–297, 1975.

GOUGH, D. I., The geophysical significance of geomagnetic variation anomalies, *Phys. Earth Planet. Inter.*, **7**, 379–388, 1973.

GREENHOUSE, J., Geomagnetic time variations on the sea floor off Southern California, Ph. D. Thesis (SIO Ref. 72–67), Univ. of Calif., San Diego, 1972.

HARVEY, R. R., Oceanic water motion derived from the measurement of the vertical electric field, Hawaii Institute of Geophysics, Univ. of Hawaii, Honolulu, 1972.

HARVEY, R. R., J. C. LARSEN, and R. MONTANER, Electric field recording of tidal currents in the Straight of Magellan, *J. Geophys. Res.*, **82**, 24, 3472–3476, 1977.

KLEIN, D. P., Magnetic variation (2–30 cpd) on Hawaii Island, and mantle electrical conductivity, Ph. D. Thesis, Univ. of Hawaii, 1975.

LARSEN, J. C., Low frequency signals (0.1–6.0 cpd) electromagnetic study of deep mantle electrical conductivity beneath the Hawaiian Islands, *Geophys. J. R. Astr. Soc.*, **43**, 17–46, 1976.

LAUNAY, L., Conductivity under the ocean: interpretation of a magnetotelluric sounding, 630 km off the California coast, *Phys. Earth Planet. Inter.*, **8**, 183–186, 1974.

MANGELSDORF, P. C., Jr., Gulf stream transport measurements using the salt bridge GEK and Loran C navigation, *Trans. Am. Geophys. Union*, **49**, 198 (abstract), 1968.

MCWILLIAMS, J. C., Maps from the mid-ocean dynamics experiment: Part I. Geostrophic streamfunction, *J. Phys. Oceanogr.*, **6**, 810–827, 1976.

MODE, *Planning Statements and Progress Report*, The MODE Scientific Council, M. I. T., Cambridge Mass., 02139, July 27, 1972.

PISTEK, P., Conductivity of ocean crust, Ph. D. Thesis, Univ. of Calif., San Diego, 1977.

POEHLS, K. A. and R. P. VON HERZEN, Electrical resistivity structure beneath the Atlantic Ocean, *Geophys. J. R. Astr. Soc.*, **47**, 331–346, 1976.

RICHARDS, M. L., Tidal signals on long submarine cables, *Ann. Geophys.*, **33**, 1/2, 177–178, 1977.

RIKITAKE, T., Global electrical conductivity of the earth, *Phys. Earth Planet. Inter.*, **7**, 245–250, 1973.

SCHMUCKER, U., Regional induction studies: a review of methods and results, *Phys. Earth Planet. Inter.*, **7**, 365–378, 1973.

SERSON, P. H., Instrumentation for induction studies on land, *Phys. Earth Planet. Inter.*, **7**, 313–322, 1973.

Atlantic Lithosphere Sounding

C. S. Cox,*[1] J. H. Filloux,*[1] D. I. Gough,*[2] J. C. Larsen,*[3]
K. A. Poehls,*[4,†] R. P. Von Herzen,*[4] and R. Winter*[5,††]

*[1]*Scripps Institution of Oceanography, La Jolla, California, U.S.A.*
*[2]*Institute of Earth and Planetary Physics, University of Alberta, Edmonton, Canada*
*[3]*NOAA/PMEL, NE, Seattle, Washington D.C., U.S.A.*
*[4]*Woods Hole Oceanographic Institution, Woods Hole, Massachusetts, U.S.A.*
*[5]*Institut fur Geophysik, Universitat Gottingen, Gottingen, F.R.G.*

(Received February 17, 1978; Revised November 17, 1979)

We have measured electromagnetic fluctuations at a number of stations on islands and on the sea floor of the Atlantic. Magnetic fields H were measured on the Bahamas, on Bermuda and at two sea floor stations. Horizontal components E of the electric field were measured on Bermuda and at three deep sea stations. The latter form a triangular array with 100 km legs on the Hatteras Plain. The age of the underlying lithosphere is estimated at 130 million years.

E at the sea floor is dominated by the Lorentz force $U \times B$ below 0.1 cpd. These fields have a typical horizontal scale less than 100 km and are unsuitable for deep electromagnetic soundings but provide limits to shallow conductivity. At higher frequencies, up to the oceanic cut off of several cycles per hour, the fields are of ionospheric origin.

The impedance tensor has been estimated at Bermuda and two sea floor stations for the ionospheric fields. On Bermuda the local influence of the island has been removed by Larsen's method for comparison with the sea floor impedances. At both sea floor stations where impedances were measured, only the component generated by eastward E and northward H is well determined. Models of lithospheric conductivity which are consistent with all measurements show diminishing conductivity below a crustal layer, then sharp rises at 70 and 250 km.

1. Introduction

The MODE-I ocean experiment of 1973 was a large cooperative experiment in physical oceanography carried out in the Sargasso Sea midway between Bermuda and the Bahama Islands. The primary aim of the experiment was to study the "mesoscale" ocean water current features which are found in this region. As a contribution to this experiment we installed electric and magnetic recorders at a number of sea floor sites and on Bermuda and on one of the Bahamas. The main purpose of these measurements was to infer the low frequency, barotropic components of water flow from their electromagnetic signatures (Cox *et al.*, 1978). The higher frequency components of the fields mea-

† Present address: Dynamics Technology Incorporated, Torrance, California, U.S.A.
†† Present address: DFVLR, Oberpfaffenhofen, F.R.G.

sured at the array of stations are suitable also for electromagnetic sounding of the crust and mantle. The impedances of the earth indicated by these oceanic and island observations and the sources of uncertainty in the impedance relationships form the principal part of this report. We also present a preliminary model of electrical conductivity consistent with the observed impedance.

2. Distribution of Stations and Periods of Observation

The locations of observations are shown in Fig. 1. The temporary island magnetic stations (Bermuda and Marsh Harbor, Bahama Island) were under the supervision of D. I. Gough. The instruments at each site were three component variographs protected

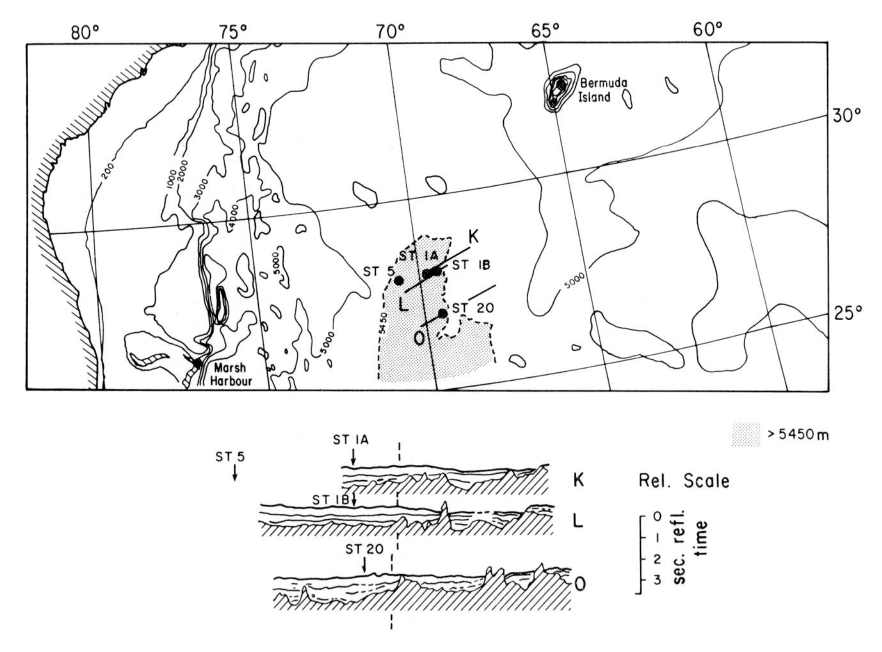

Fig. 1. Map shows relative positions of stations. Island stations were on Bermuda and at Marsh Harbor, Great Abaco. Stations 1A, 1B, 5, and 20 were at the sea floor in the Hatteras Abyssal Plain (shaded area). Acoustic reflection sections are available along the tracks marked K, L, O. The diagrams below the map show the sections with acoustic basement indicated by cross hatching. Note that although the basement penetrates the sediments as mountain ridges to the east, the stations are on simple plane parallel sedimentary structures. The sections were provided by T. Ewing.

from thermal fluctuations by burial of the observation chambers one meter in the ground. J. C. Larsen conducted electric field observations on Bermuda by measuring the electric potential differences between three points in Castle Harbor. R. P. Von Herzen and K. Poehls installed a three component magnetic variograph on the sea floor first at MODE Station 5 (March–May) and later at Station 20 (May–July). J. H. Filloux and C. S. Cox installed sea floor electric field recorders of Filloux's design (FILLOUX, 1974) at each of the MODE Stations 1, 5, and 20. Each of these deep sea electric recorders registers the two components of electric field parallel to the sea bed. During the first period of observation (March–May) a fourth electric recording meter was installed for comparison purposes also at Station 1. During this period the two recorders at Station 1 (actually separated by about 10 km from one another) are referenced as Stations 1A and 1B.

The location of stations and periods of observation are summarized in Table 1. In addition to the temporary recording sites listed above, two nearby permanent geomagnetic observatories are maintained by the U. S. National Ocean and Atmospheric Administration: Fredricksburg, Virginia and San Juan, Puerto Rico.

Table 1. *E/M* observations conducted in MODE.

Observation*	Location			Recording dates
B	Bermuda, Castle Harbor			16 Mar–11 July
B	Sta 5[†]	27°50′N	70°40′W	15 Mar–23 Mar
				24 Mar–14 May
B	Sta 20	27°09′N	69°31′W	17 May–11 July
B	Great Abaco, Bahamas			18 Mar– 9 July
E	Bermuda			15 Mar–11 July
E	Sta 1A	27°57.2′N	69°39.9′W	24 Mar–14 May
				17 May– 7 July
E	Sta 1B	27°58.7′N	69°33.7′W	16 Mar–14 May
E	Sta 5[†]	27°50′N	70°40′W	15 Mar–22 Mar
				23 Mar–15 May
				15 May– 6 July
E	Sta 20	27°08′N	69°32′W	20 Mar–10 May
				16 May– 8 July

* *B* indicates 3 components of magnetic variations. *E* indicates 2 horizontal components of electric variations.

[†] Initial *B* and *E* observations at Sta 5 were at 27°50′N, 70°50′W (15–23 Mar).

3. General Character of Electromagnetic Fields on Islands and the Sea Bed

Selected electric and magnetic fields are presented for comparison in Fig. 2. At very low frequencies (say, less than 0.1 cpd) the sea floor electric fields show strong activity which has no appreciable counterpart in the magnetic fields. The comparison electric recordings 1A and 1B exhibit the same oscillations in this frequency range except for a steady drift which has been tentatively identified as the result of electrolytic corrosion on the Station 1B recorder (see next section).

The low frequency parts of the electric field at Stations 5 and 20 differ significantly from Station 1 but are each highly active in this frequency range.

The origin of this activity is primarily induction of horizontal electric components by motion of ocean water through the earth's magnetic field:

$$\boldsymbol{E} = -\boldsymbol{u} \times \boldsymbol{B}_0 + \boldsymbol{j}/\sigma_{\mathrm{w}} \tag{3.1}$$

where $\boldsymbol{u}$ is the water current, $\boldsymbol{j}$ the electric current, σ_{w} the conductivity of seawater, and $\boldsymbol{B}_0$ the geomagnetic induction. Under certain circumstances the horizontal electric

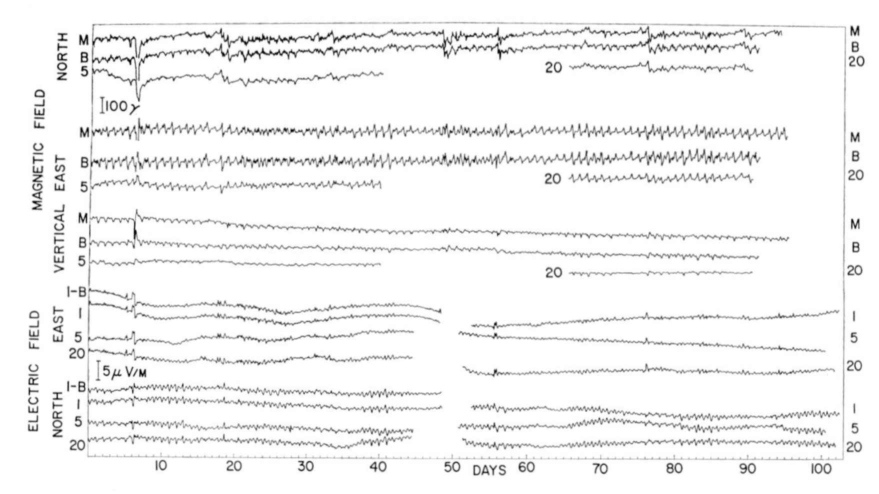

Fig. 2. Selected electromagnetic fields. M and B refer to magnetic observations at Marsh Harbor and Bermuda respectively. Numerals refer to sea floor stations as shown in Fig. 1 (note that Station 1 is the same as Station 1A). The time scale begins (day 0) 1973, 26 March 14 hr UT.

current integrated vertically through the ocean is negligible and one can relate the horizontal components of the electric field to the EMF induced by the horizontal barotropic water current $\boldsymbol{U}$ by

$$\boldsymbol{E} = -\boldsymbol{U} \times \hat{\boldsymbol{e}}_3 B_3 \tag{3.2}$$

in which $\hat{\boldsymbol{e}}_3$ is a unit vertical vector and B_3 is the vertical component of the geomagnetic field. A comparison of $\boldsymbol{E}$ and an estimate of $\boldsymbol{U}$ for Station 1 derived by FREELAND and GOULD (1976) is shown in Fig. 3. The comparison of these completely independent data sets is surprisingly close, indicating that the conditions necessary for the validity of Eq. (3.2) are well satisfied. We may use this observation to set an upper limit to the conducting properties of the crust in immediate contact with the ocean.

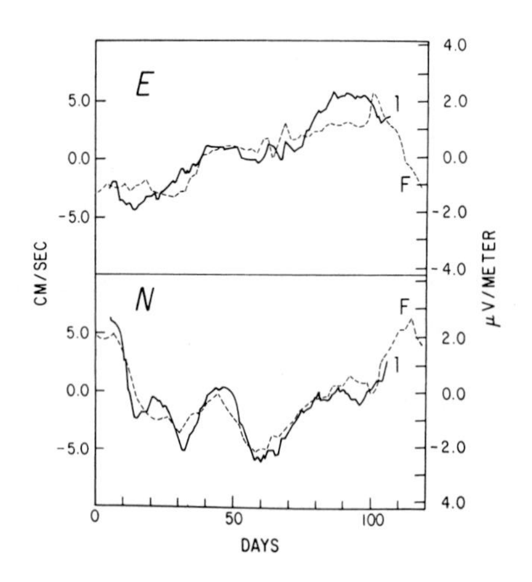

Fig. 3. Comparison of low frequency filtered electric filed at Station 1 (1) components with $\boldsymbol{U}$ from Freeland and Gould (F). Upper panel: East component of $\boldsymbol{U}$ and south electric field. Lower panel: North component of $\boldsymbol{U}$ and east electric field.

The argument may be summarized as follows. The horizontal electric field tends to drive leakage electric current into the sea bed. The resistive voltage drop of this electric current reduces the left side of Eq. (3.2) relative to the right side. The close similarity of $\boldsymbol{E}$ to the EMF as indicated visually in Fig. 3 implies that the leakage current is negligible, i.e., that the impedance of the sea bed is high compared to the load impedance of the sea floor. We must estimate this ratio. Let the conductivities of sea water and underlying rock be σ_w and σ_r respectively. If σ_r is homogeneous and isotropic and the depth of the ocean is h, the reduction factor of the left to right sides of Eq. (3.2) can be shown to be

$$E/E_0 = [1+\sigma_r/(kh\sigma_w)]^{-1} \tag{3.3}$$

for water flow distributed in a sinusoidal pattern with stream function Ψ according to

$$\begin{aligned} \boldsymbol{U} &= \nabla \times (\hat{\boldsymbol{e}}_s \Psi) \\ \Psi(\boldsymbol{r}) &= \mathrm{Re}\,[\psi \exp(i\boldsymbol{k}\cdot\boldsymbol{r})]\,, \end{aligned} \tag{3.4}$$

where $\boldsymbol{k}$ is the wavenumber of the flow distribution and $\boldsymbol{r}$ is the horizontal position vector.

We note that electric current flow is a static effect: the frequency of the MODE eddies is far too low to have an appreciable inductive effect close to the sea bed.

Freeland and Gould have found a statistical representation of the spatial distribution of water currents in the MODE experiment which is estimated to be valid for the barotropic currents. The wavenumber spectrum is assumed to be isotropic and equivalent to a spectrum of streamfunction

$$S_\psi(\boldsymbol{k}) = Ca^5(a^2+k^2)^{-7/2}\,, \tag{3.5}$$

where C is an intensity factor and $a \simeq .032\ \mathrm{km}^{-1}$ establishes the spatial character of the flow. In the absence of electric leakage into the sea bed the spectrum of electric field intensity would be that of the right side of Eq. (3.2):

$$S_{E_0}(\boldsymbol{k}) = (kB_3)^2 S_\psi(\boldsymbol{k})\,. \tag{3.6}$$

This relation follows from the definition of the streamfunction, Eq. (3.4). With allowance Eq. (3.3) for leakage to the sea bed the spectrum of the left side is

$$S_E(\boldsymbol{k}) = (kB_3)^2[1+\sigma_r/(kh\sigma_w)]^{-2} S_\psi(\boldsymbol{k})\,. \tag{3.7}$$

After combination of Eq. (3.5) with Eqs. (3.6) and (3.7) we find the ratio

$$\langle R\rangle = \left[\int_{-\pi}^{\pi}\int_0^\infty kS_E(\boldsymbol{k})\mathrm{d}k\mathrm{d}\theta\right] \Big/ \int_{-\pi}^{\pi}\int_0^\infty kS_{E_0}(\boldsymbol{k})\mathrm{d}k\mathrm{d}\theta$$

for the mean square of left to right sides of Eq. (3.2). After combination of Eqs. (3.5) and (3.7) we find

$$\langle R\rangle = 1-2.0f+o(f^2) \tag{3.8}$$

where $f=\sigma_r/(ha\sigma_w)$. The visual impression of similarity between the solid and dashed lines in Fig. 3 suggests that the ratio is greater than 0.5. Hence $2.0f<0.5$ and therefore $\sigma_r<0.29$ S/m. This upper limit is effective only in the rocks near the sea floor, to a depth of order $a^{-1}=30$ km. Furthermore, it is not directly comparable to conductivity derived from magnetotelluric work because the electric current flow, lying in the vertical plane, may respond primarily to the vertical conductivity, whereas in magnetotellurics the electric current flow is horizontal. One may also object to the use of statistical expectation for a comparison which has a short duration relative to the time scales of the mesoscale flow. Nevertheless our method indicates the essential roles of the spatial scale a and rock conductivity in controlling sea floor leakage from ocean EMF's induced by barotropic flow.

In the usual electromagnetic sounding methods one assumes that the fields are largely of the transverse electric (TE) mode and that the horizontal scale of the sources is large compared to the depths of penetration within the earth. If return flow occurs within the solid rocks below the ocean, the oceanic fields induced by mesoscale water motions must generate electric currents in the vertical plane; these represent fields of the transverse magnetic (TM) polarization. Furthermore, they have a short horizontal scale. The magnetic fields, if detectable at all, will be very weak and the impedance E/H will be much larger than for the TE mode (see Fig. 2 of Cox *et al.*, 1970 for examples of the influence of mode and horizontal scale on the impedances).

At frequencies between 0.5 cpd and 10 cph there is a close resemblance between the electric fields at all ocean stations (Figs. 4 and 5) and evident synchronism of disturbances on recording of electric and magnetic fields. The principal source of these disturbances appears to lie in the ionosphere since the signal appears in both island recordings as well as at sea. Secondary, although less obvious sources lie in oceanic tides, internal waves, and turbulence.

At frequencies above 10 cph the oceanic electric fields still have recognizable ac-

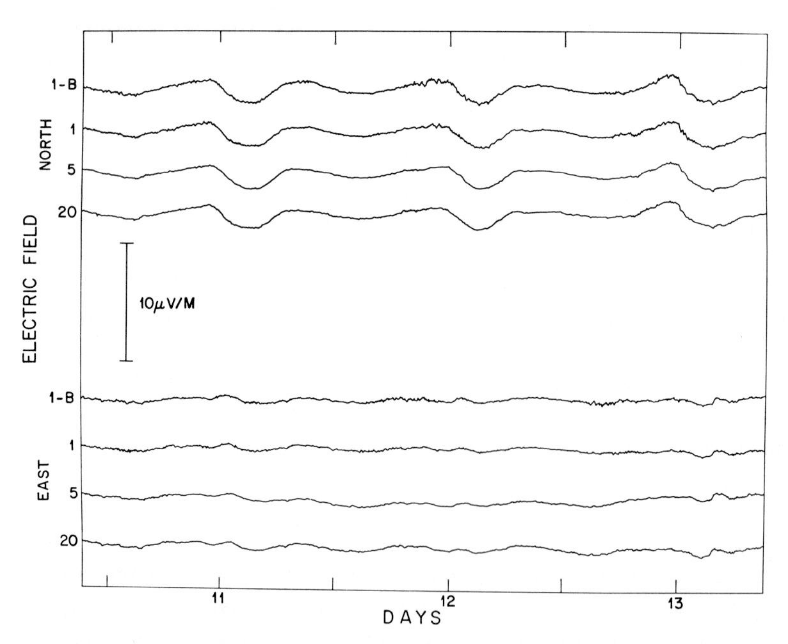

Fig. 4. Horizontal electric fields at the sea floor during a magnetically quiet time. The time scale is that of Fig. 2.

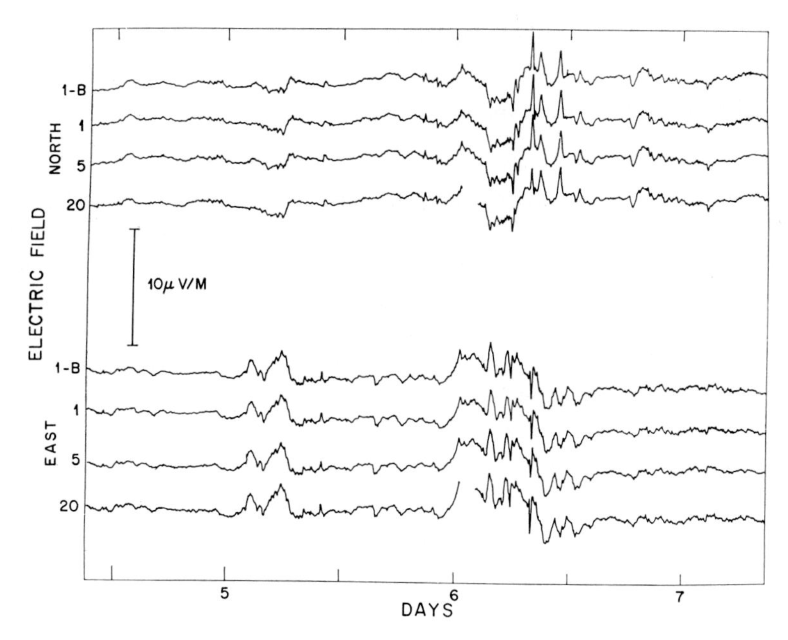

Fig. 5. Similar to Fig. 4 but during a magnetic storm.

tivity but they are no longer coherent from station to station, nor is the magnetic signature above recorder noise. We have tentatively assigned these fluctuations to induction processes involving turbulence at the sea bed.

The electrical properties of the oceanic signals are summarized in the spectrum (Fig. 6). The lowest frequencies are dominated by induction derived from mesoscale water motions. There is a gap between 0.1 and 0.5 cpd followed by the region dominated by ionospheric sources. Standing above this region are the sharp lines of ocean and

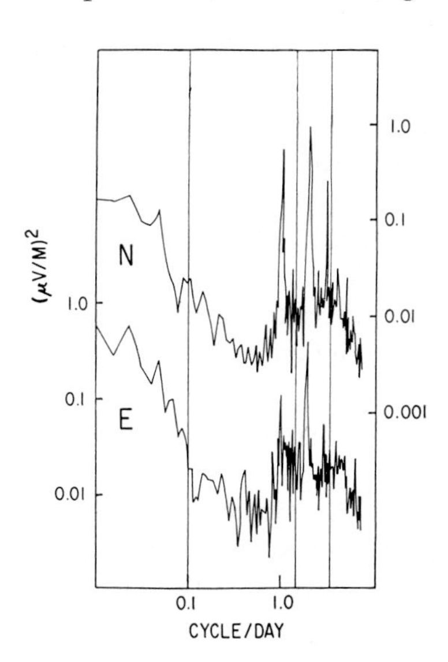

Fig. 6. Sea floor electric spectra averaged over all stations and for all available data. The quantity plotted is the frequency times the spectrum. The valley at about 0.5 cpd occurs between the low frequency part dominated by the induction from water motions and the high frequency part dominated by ionospheric signals.

atmospheric tides, and the solar daily variation.

4. Data Processing and Uncertainties

The data from all the temporary instruments are subject to errors which must be recognized to avoid erroneous interpretation. The island magnetic stations are stable except for thermal drift. Burial of the instruments reduced the thermal noise, but there are transients which occurred at the beginning of recording and after servicing of the instruments. The Castle Harbor electric recordings are influenced by unwanted interferences driven by local water currents in the harbor and temperature and salinity fluctuations near the electrodes. The oceanic magnetometers utilized a new, thin film sensor which exhibited temporal drift and high frequency noise. These errors seem to be most serious below 1 cpd and above 3 cph respectively.

The sea floor electric recorders were of a new type (FILLOUX, 1974) which record horizontal electric fields through two orthogonal salt bridges in the horizontal plane. The effective separation over which the field components are measured is 6 m. The salt bridges through which the field is measured are mounted on the recorder body 2.5 m above the sea floor.

One of the sea floor electric instruments was subject to evident corrosion caused by failure of insulation on a radio antenna. This introduced a steady drift in the record. The other instruments did not appear to be significantly corroded but unrecognized causes of drift cannot be excluded. A more serious problem was that several of the tape drive mechanisms on digital recorders slipped so that it was very difficult to read the tapes and some of the readings were garbled. An extensive data correction scheme, depending on comparison of electric recordings of all stations was required to clean up the records. Station 1B was the most seriously affected, while the recordings in the second period (May–July, all stations) were least affected. The electric readings were made at 16 second intervals. Clean-up consisted of replacement of garbled data by a spline fit over adjacent accepted data followed by a trapezoidally weighted smoothing, and resampling at 64 second intervals.

Another serious problem with the electric recorders was that the clocks had large and unknown rate errors. It was found possible to synchronize the independently recorded electric and magnetic records by calculation of the coherency and phase shift between the two for successive two day sections of the records. It was possible to find usable phase shifts at frequencies up to 10 cph. A simple translation of the origin of time and uniform stretching of the time coordinate was successful in reducing the time uncertainty to ± 20 sec. The correction involves the assumption that the phase lead of E relative to H at frequencies close to 10 cph is 45 degrees.

There are uncertainties related to the orientation of the sea floor magnetic and electric recorders. Every effort was made to insure that the magnetometer was not tilted by variable dynamic water forces on its mounting frame, but there is no certain method of detecting this source of noise. The electric recorders nominally record the two components of electric field parallel to the sea floor on which they rest. The tilt of the recorders was recorded and found to be steady and less than 5° in all cases. The component of electric field normal to the sea bed has a large source in the field induced by local eastward water currents cutting the horizontal northward component of the geomagnetic field. This source probably has a maximum in the frequency range of internal waves ($1\text{ cpd}<f<4\text{ cpd}$). It is quite possible that a few percent of this signal may have contaminated the electric measurements because of imperfect parallelism between the voltage pickup system and the sea floor.

A final source of uncertainty is that the azimuths of horizontal orientation of the sea floor recorders were determined separately (by recording magnetic compasses) and are subject to errors. This leads to uncertainty in the "distortion tensor" which differentiates a scalar from a tensor impedance relation (next section).

5. The TE Impedance at the Sea Floor

We have analyzed 25 day sections of data recorded at Station 20 in June–July and at Station 5 in March–April as representative of the most carefully cleaned and certain data. A preliminary analysis was reported by WINTER (1976). The response functions between horizontal components of E and B can be represented in two forms:

$$\begin{aligned} E_1&=Z_{11}B_1+Z_{12}B_2+N_1^E \\ E_2&=Z_{21}B_1+Z_{22}B_2+N_2^E \end{aligned} \tag{5.1}$$

$$\begin{aligned} B_1&=A_{11}E_1+A_{12}E_2N_1^B \\ B_2&=A_{21}E_1+A_{22}E_2+N_2^B\,. \end{aligned} \tag{5.2}$$

In our notation the axes (1, 2, 3) point respectively east, north (geographic), and upward.

The two forms (Eqs. (5.1) and (5.2)) represent the two extremes in which all the incoherent noise N is ascribed respectively to the electric components or the magnetic components. Unless the coherence between electric and magnetic oscillations is perfect (in which case $N^E=N^B=0$) it will not be found that the response functions are equivalent. To keep the distinction clear between the two methods of determining them while working with the response in the form E/B we shall write

$$[Q_{ij}]=[A_{ij}]^{-1} \tag{5.3}$$

for the "impedance" derived as the inverse of the "admittance" $[A]$. In a general way we see that Eq. (5.1) underestimates the true impedance because some of the recorded electric signals are assumed to be noise, while Eq. (5.2) overestimates it. Our purpose is to find the allowable range within which the true TE response function for large scale fields lies. Because of the wide range of possible noise sources both of a geophysical and instrumental nature we are unable to ascribe the true noise to either extreme. Our tentative assumption is that none of the noise sources introduces a systematic influence on the response functions, but only reduces the coherency of E and B. This assumption will need modification if it turns out that the ionospheric sources are not of sufficiently large scale, or if oceanic water motions introduce a systematic impedance. For the present we neglect such complications.

We have utilized the mid frequency band $0.1\text{ cph}<f<3.5\text{ cph}$ because this range is in the continuum away from tidal lines and in the frequency range where we expect the

Table 2. Response tensor for Station 20 (no adjustments).

f (cph)	Z_{11}		Z_{12}		Z_{21}		Z_{22}	
	Real	Imag	Real	Imag	Real	Imag	Real	Imag
.104	−.00302	.00024	−.01421	.03847	.02351	−.00966	−.00897	.00797
.187	−.00350	.00486	−.03059	.04282	.01391	−.02027	−.00284	.00657
.271	−.00527	.01337	−.03277	.04743	.00791	−.01433	.00048	.02163
.354	−.01121	.01545	−.03980	.05425	.01239	−.01608	−.00407	.01610
.521	−.01604	.02156	−.04992	.06869	.01057	−.01631	−.00961	.01999
.729	−.02062	.03148	−.06695	.08159	.01131	−.01394	−.02135	.03001
.937	−.01837	.02792	−.06800	.08830	.01965	−.02070	−.02560	.03496
1.500	−.02326	.04127	−.08291	.10111	.02264	−.02075	−.03128	.04654
2.500	−.03894	.03378	−.08570	.08757	.02330	−.03410	−.03776	.05026
3.500	−.04357	.02407	−.07636	.07290	.02783	−.05160	−.03169	.03930
4.500	−.04808	.02387	−.05775	.05188	.02024	−.04778	−.03508	.04085
5.500	−.04037	.01881	−.05958	.03455	.02410	−.04051	−.01932	.02446
6.500	−.03623	.02072	−.04908	.03369	.01559	−.03582	−.02072	.02159
7.500	−.02796	.01379	−.04140	.02714	.00733	−.02970	−.02156	.01287
8.500	−.02575	.01287	−.02888	.01290	.01108	−.02642	−.01239	.00626
9.500	−.02106	.00785	−.02885	.00812	.00756	−.01861	−.01404	.00698
f (cph)	Q_{11}		Q_{12}		Q_{21}		Q_{22}	
.104	.00110	.00089	−.01460	.04793	.03893	−.01209	−.00492	.01224
.187	−.00656	−.00008	−.03810	.05312	.02700	−.04273	−.01328	.00786
.271	−.00715	.01124	−.03776	.05710	.01831	−.04512	−.00304	.03712
.354	−.00972	.01450	−.04622	.06564	.03687	−.04639	−.00254	.03741
.521	−.01731	.01787	−.05337	.08039	.02792	−.03915	−.01086	.04106
.729	−.01596	.03116	−.07669	.10107	.04215	−.03676	−.02863	.08302
.937	−.02141	.02727	−.07403	.10097	.04374	−.04186	−.02384	.05917
1.500	−.01784	.05294	−.11474	.14694	.05119	−.03328	−.04546	.09878
2.500	−.04183	.07849	−.20950	.20236	.07173	−.05859	−.08378	.16087
3.500	−.04653	.07761	−.26530	.22598	.08502	−.09673	−.08641	.16984
4.500	−.10853	.10988	−.32377	.24098	.07674	−.11028	−.14828	.20732
5.500	−.09621	.02344	−.40682	.26833	.10576	−.17384	−.19400	.22725
6.500	−.12951	.07002	−.38609	.26470	.09128	−.18940	−.19843	.23523
7.500	−.12555	.12388	−.45324	.23284	.11174	−.23923	−.21575	.23470
8.500	−.11240	.09253	−.65995	.23990	.11601	−.20053	−.30820	.26371
9.500	−.18707	.09348	−.53176	.11954	.11493	−.21658	−.28959	.20718

Units are μV/m nT (=km/s).

maximum signal to noise ratio of large scale ionospheric sources relative to oceanic motional sources. Details of the computation of the tensors [Z] and [A] and their uncertainties will be given elsewhere. Results for Station 20 are summarized in Table 2. The spectra (Fig. 7) show separately the whole spectrum of E and B and the parts of E incoherent with B. It is clear that the north component of E is both weaker than the east component and subject to more interference from noise sources at frequencies below 1.5 cph. As a consequence the response functions linking the north component of E to B are much less certain than those linking the east component. At higher frequencies the situation is reversed.

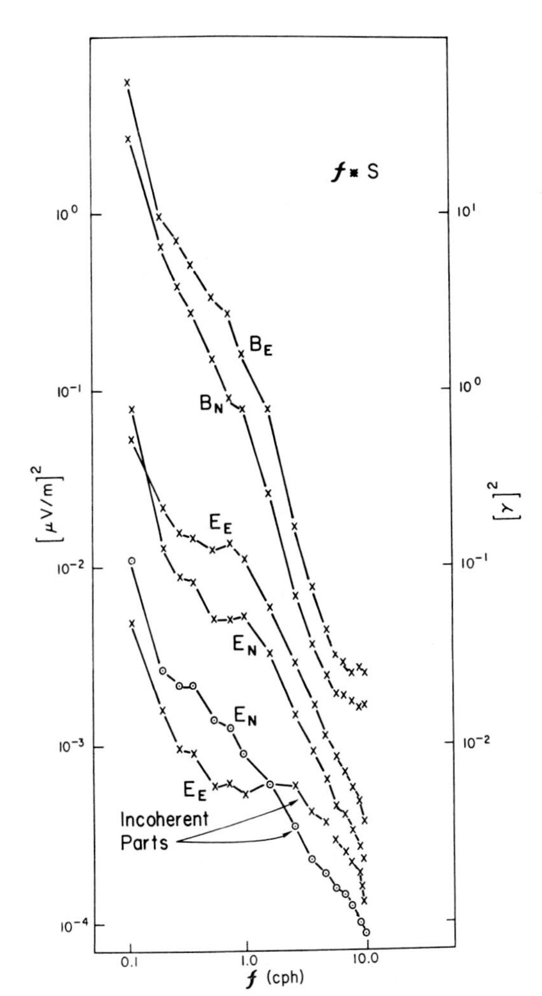

Fig. 7. Spectra of the horizontal components of E and B during a 25 day, magnetically disturbed time in June 1973 at Station 20. The quantity plotted is the frequency times the spectrum. The parts of E incoherent with B are noted.

The response functions are skewed and anisotropic. It is conceivable that these features are the result of local conditions near the electric recorders on the sea floor. For example, if poorly conducting basement rock were to penetrate the sediments close to the recorder, the electric field would be locally distorted because electric currents flowing in the sediments would be forced into the sea water by the poorly conducting rocks. If the deep structure is horizontally isotropic but the surface electric field is distorted by a shallow and poorly conducting structure, the relation between observed E and B fields should be (LARSEN, 1975),

$$\begin{pmatrix} E_1 \\ E_2 \end{pmatrix} = \bar{Z} \begin{pmatrix} A+C & -1+B \\ 1+B & A-C \end{pmatrix} \begin{pmatrix} B_1 \\ B_2 \end{pmatrix} \tag{5.4}$$

in which the distortion elements A, B, C should be real and independent of frequency while $\bar{Z}(f)$ contains all the deep sounding information except for an unknown scale factor D which is also a part of the local distortion of E. In Fig. 8 we show each of

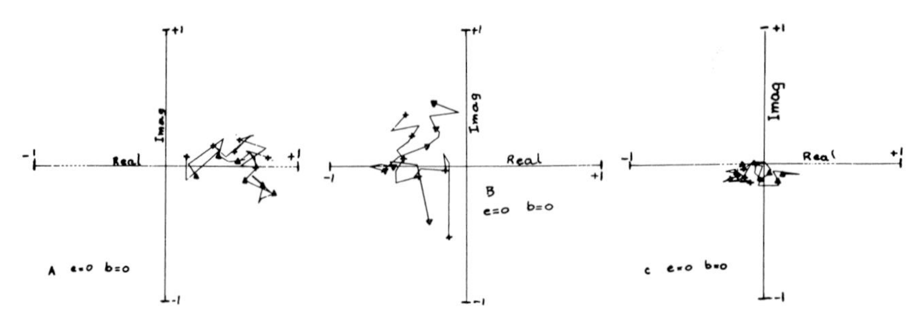

Fig. 8. Real and imaginary parts of A, B, C for Station 20.

A, B, and C derived from the impedance ratios

$$A=(Z_{11}+Z_{22})/(Z_{12}-Z_{21})\,, \quad B=-(Z_{12}+Z_{21})/(Z_{12}-Z_{21})\,,$$

$$C=(Z_{11}-Z_{22})/(Z_{12}-Z_{21})\,,$$

and also from the corresponding expressions with Q. The points are connected by straight lines in sequential order in frequency. Triangles on every third point indicate A, B, and C values derived from Z while pluses on every third point of A, B, and C values are derived from Q.

If A, B, C were indeed real quantities another interpretation would be possible which does not involve distortions of the electric field. An apparent term A will be produced if there is an error in orientation of the electric or magnetic recorders. In this case we can eliminate the apparent A by a differential rotation between the coordinate systems through angles e for the electric and b for the magnetic vectors respectively, such that $\tan(e-b)=A$. The remaining terms B and C can be regarded as an (unskewed) anisotropy in the conducting matter below the recorders. The term C can be removed by rotating to principal axes according to $\tan(e+b)=-C/B$. The angles are measured counterclockwise (looking down). This development shows that the term A of the distortion tensor can be caused by an error of orientation of either recorder, or a uniform, frequency independent rotation of either vector field. The deviation of B and C from zero implies either a stretching distortion of one of the fields along the principal axis or simple anisotropy of conductivity within the underlying rocks. After deskewing and rotation to principal axes the modified transfer tensor takes the form

$$\bar{Z}'\begin{pmatrix} 0 & -1+B' \\ 1+B' & 0 \end{pmatrix} \tag{5.5}$$

where

$$B'=\pm\sqrt{B^2+C^2}/\sqrt{1+A^2} \tag{5.6}$$

is a measure of the anisotropy and

$$\bar{Z}'=\pm\bar{Z}\sqrt{1+A^2} \tag{5.7}$$

contains the phase and frequency information of the magnetotelluric sounding.

It appears from Fig. 8 that A and B do have systematic variations with frequency, and that they have non zero phases. Therefore neither of the simple models considered above are completely adequate. Furthermore the uniformity of ocean depth in the vicinity of Station 20 and the lack of shallow basement irregularities under the station (Fig. 1) argue forcefully against any local distortion of the electric field. Nevertheless we have assumed that at least some of the skew is caused by errors in determination of instrument azimuths. Consequently we have attempted the deskewing and rotation to principal axes by choice of e and b according to

$$\tan [2(e-b)] = 2 \text{ Re } (A)/(1-|A|^2)$$

$$\tan [2(e+b)] = -2 \text{ Re } (BC^*)/(|B|^2-|C|^2)$$

which minimize the deskewed and rotated values of A and C respectively at a fixed frequency. (The asterisk indicates the complex conjugate.) The new values of the distortion, A', B', and C' are plotted in Fig. 9 for a good choice $e=20°$ and $b=-5°$ for Station 20. The corresponding response functions are listed in Table 3.

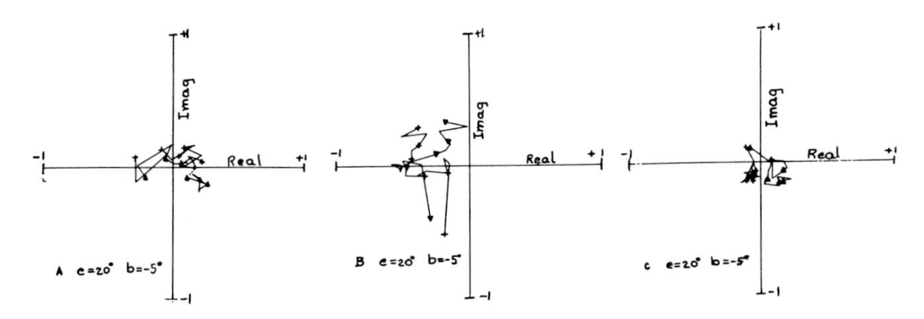

Fig. 9. Similar to Fig. 8 except that the response tensor has been deskewed and rotated to principal axes ($e=20°$, $b=-5°$).

The distortion elements are nondimensional and to be compared with unity (the undistorted off-diagonal term in Eq. (5.4)). The frequency dependence of C' is quite small while that of A' and B' is not *very* large. However the latter do seem to have a systematic and complex frequency effect. This seems to favor some large scale and probably buried source of anisotropy and skew. There do not seem to be any local variations in sediment thickness which could produce the effects (Fig. 1). We do not at present think that it can be caused by local variations of water depth because the depth varies by less than 10% within 200 km of the station and furthermore Station 20, lying as it does in the bottom of a north-south oriented basin, should have weaker east-west electric fields rather than *stronger* as observed.

The analysis for Station 5 is essentially the same as that performed on the Station 20 data except for the method of smoothing. Station 20 utilized band-averaging within the 25 day data record. Station 5 analysis was performed on the whole 25 day record only for periods longer than 2 hr. Subsections of the 25 day record with large amplitude short period activity were chosen to determine response functions at periods shorter than 2 hr. The results from these subsections were then ensemble averaged.

Response functions for Station 5 are presented in Table 3. They have been deskewed and rotated to principal axes. The best values of $e=11°$ and $b=-2°$ yield a principal

Table 3. Response functions for ocean stations after deskewing and rotating to principal axes.

Station 5 $e=11°$ $b=-2°$

f (cph)	Z_{12} Real	Z_{12} Imag	Z_{21} Real	Z_{21} Imag	Q_{12} Real	Q_{12} Imag	Q_{21} Real	Q_{21} Imag
.17	− 8.5	11.3	17.7	26.3	− 24.	43.	31.5	−43.1
.22	−22.0	32.0	8.2	17.7	− 29.	46.	27.8	−66.0
.42	−34.6	42.6	9.6	12.9	− 36.	54.	30.7	−44.5
.67	−45.4	42.1	18.0	13.4	− 55.	66.	36.3	−25.4
1.05	−52.8	57.7	25.2	20.2	− 68.	91.	43.3	−28.5
1.67	−65.3	54.5	21.3	24.2	− 86.	98.	48.3	−35.8
2.61	−71.8	37.1	12.3	34.2	−118.	112.	50.5	−54.5
4.29	−68.4	24.3	7.1	42.2	−212.	151.	58.5	−89.9

Station 20 $e=20°$ $b=-5°$

f (cph)	Z_{12} Real	Z_{12} Imag	Z_{21} Real	Z_{21} Imag	Q_{12} Real	Q_{12} Imag	Q_{21} Real	Q_{21} Imag
.104	−15.8	38.5	23.3	− 8.6	14.1	− 48.7	36.0	− 11.2
.187	−29.5	42.1	13.5	−19.9	39.9	− 51.1	27.5	− 39.0
.271	−30.7	52.4	8.2	−18.3	36.4	− 65.7	18.7	− 47.4
.354	−39.2	57.0	14.6	−20.0	43.9	− 74.0	36.7	− 49.5
.521	−51.0	72.4	14.7	−22.2	54.3	− 89.3	31.3	− 43.7
.729	−71.3	88.8	17.4	−23.8	81.6	−124.4	45.0	− 48.8
.938	−73.3	96.2	24.7	−29.1	77.9	−115.7	48.0	− 50.3
1.50	−89.5	113.3	29.2	−34.3	122.8	−174.6	54.3	− 52.9
2.50	−95.6	100.9	35.6	−44.9	225.9	−248.9	82.0	− 88.7
3.50	−85.0	82.1	41.2	−57.5	279.1	−272.9	94.6	−124.2
4.50	−69.4	63.0	36.5	−54.7	360.2	−301.9	111.3	−150.5
5.50	−64.9	41.0	36.1	−45.3	451.7	−325.4	135.5	−181.3
6.50	−55.5	39.5	27.2	−41.4	436.9	−328.0	134.4	−212.5
7.50	−48.2	30.0	16.9	−32.7	504.7	−300.9	151.5	−278.4
8.50	−33.0	14.5	19.3	−29.3	528.6	−316.0	152.5	−233.7
9.50	−33.3	10.1	14.5	−20.4	608.4	−183.7	179.2	−248.0

direction of approximately 10° west of geographic north, similar to that found for Station 20. The response functions again exhibit significant anisotropy with the easy conductivity approximately north-south.

6. Conductivity Models for Sea Floor Stations

The origin of the skew term A in Eq. (5.4) is uncertain. A considerable part could possibly be the result of misinformation on the orientation of the recorders on the sea bed. Accordingly, we have adopted the "deskewed" response function of Table 3 for comparison with models. The coordinates are also rotated as described above to reduce the anisotropy to principal axes. If all the hypothetical orientation error is in the electric recorder, the axis of minimum impedance is along the geographic azimuth 340°. If the orientation error is in the magnetic recorder the minimum impedance is along azimuth

$+5°$. The adjusted response functions contain residual skew, which we disregard, and a large anisotropic term B'.

An instructive presentation of the response function is in the Schmucker-Weidelt coordinates σ^*, z^*. In our notation

$$\sigma^* = \omega/(2\mu[\mathrm{Re}\,(Z)]^2)\,, \quad z^* = |\mathrm{Im}\,(Z)|/\omega\,.$$

These quantities can be determined from either of the response functions Z or Q and for either of the two principal directions, Fig. 10. The figure shows first of all (for Station 20) a rough estimate of the conductivity σ^* as a function of depth z^*. By the discrepancy between the two methods of computation it indicates the usefulness of the data. From the figure we notice that the average conductivity in the east-west direction is appreciably less than in the north-south direction. The conductivity will be most reliably determined in the east-west direction, and in the depth range of 50 to 150 km. All these features will have counterparts in our more complete model studies.

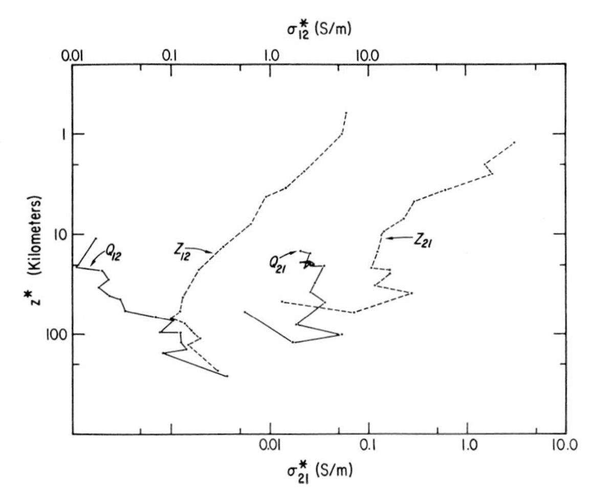

Fig. 10. The response functions from Station 20 deskewed and rotated to principal axes represented in Schmucker-Weidelt coordinates. Left side: Data from Z_{12} and Q_{12}. Right side: Data from Z_{21} and Q_{21}.

A criterion for consistency of the response for a one dimensional conductivity structure is that z^* should diminish monotonically with increasing frequency. With the notable exception of the lowest frequency datum of Q_{21} and Z_{21}, and the poorly determined, highest frequency data of Q_{21}, this consistency is observed.

We next attempt to fit our data with one dimensional models in which the conductivity lies in homogeneous but anisotropic horizontal layers. It is required that the direction of anisotropy be the same for all layers. A model of this class will be considered acceptable provided its calculated response function Z falls approximately between the two extremes Z and Q (Eqs. (5.1)–(5.3)). We allow for anisotropy by considering independent models respectively for Z_{12}, Q_{12}, and Z_{21}, Q_{21} in the rotated frame of reference.

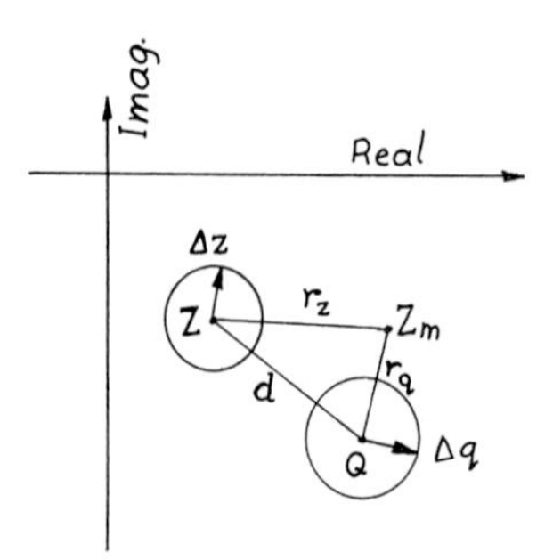

Fig. 11. The area of uncertainty of the response function for a single frequency band is shown on the complex plane. The measured impedance values which assign all incoherent noise to E and B respectively are shown at Z and Q respectively. The expected uncertainties of the measured values are shown by the circles surrounding the impedances. The fit of a model to the observed data is related to the location Z_{m} of the model impedance relative to the observed impedances.

For each model we calculate a figure of merit F indicating its acceptability. The construction of F is illustrated in Fig. 11. Within each frequency band the response estimates Z, Q, and their "circles of confusion" with radii Δz, Δq are known. In general terms the circles define those regions within which the true values of Z or Q should lie (with defined probability). In our calculation we adopt an 80% probability as an appropriate criterion. According to our hypothesis the true response function should lie in the cigar shaped area including and joining the two circles. A normalized measure of fit (indexed to frequency band by subscript i) is the ratio $f_i=(r_{zi}+r_{qi})/d_i$. The figure of merit is defined as

$$F=N^{-1}\sum_{i=1}^{N} f_i$$

where N is the number of frequency bands considered.

In order to define an acceptable model we should know the distribution of F to be expected when there are random statistical errors in the estimates of Z and Q caused by a lack of coherence between observed electric and magnetic fields. The statistics have not been worked out, but we find a useful indicator to be

$$\Delta F=N\{\sum_{i=1}^{N}[d/(\Delta z+\Delta q)]_i\}^{-1}\,.$$

Some numerical experiments lead us to believe that a reasonable estimate for an acceptable figure of merit is

$$F<1+0.4\Delta F$$

for the same uncertainty to be attached to the model as is associated with the circles of confusion whose radii are Δz and Δq.

The best defined models are those derived from Z_{12}, Q_{12}, i.e., from E approximately east, B north. Three features of the models are well established (Fig. 12). (1) There is evidence for a superficial conductivity layer, the thickness of which is poorly determined.

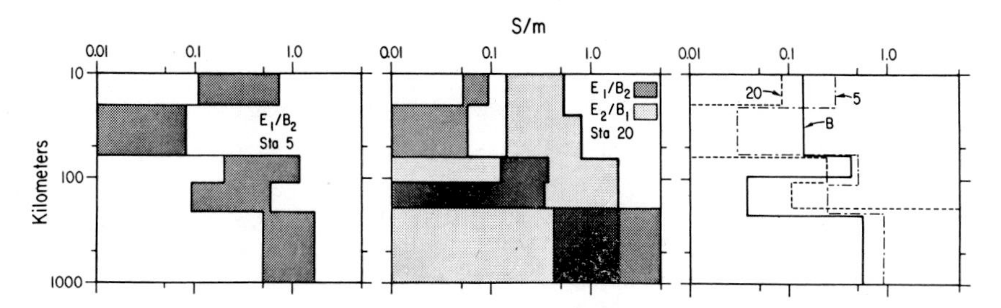

Fig. 12. Acceptable conductivity models for Station 5 (left), Station 20 (center), and Bermuda (right). The shaded areas indicate uncertainty of the sea floor models. The best fitting models derived from E_1/B_2 at the two sea floor stations are shown superimposed on the Bermuda model.

We have shown this layer as 20 km thick (and the corresponding uncertainty of conductivity), but only the integrated conductivity is established. (2) There is a sharp rise of conductivity at 75 ± 10 km. (3) There is another sharp rise at roughly 220 km. At intermediate depths (between 20 and 75 km and between 100 and 200 km) the conductivity is "invisible" because these layers are hidden by the higher conducting layers above.

The models derived from Z_{21}, Q_{21} (E north, B east) are poorly defined because the superficial layer is so highly conducting that deeper layers provide little influence on the response functions. Except for this highly conducting upper layer it is entirely possible that the conductivity distribution is the same for both polarizations. Thus we find evidence for large anisotropy (with easy conductance approximately north) only in the uppermost layer. The conductivity is surprisingly high. If distributed through a 20 km layer it must be larger than 0.3 S/m in the easy direction.

7. Conductivity below Bermuda

The input data consist of magnetic and electric oscillations observed on Bermuda. It is reasonable to ascribe all the noise to E as in Eq. (5.1) because the electric field is subject to large noises generated by local water movements, while the magnetic field is free from local disturbances.

Our method of analysis is that of LARSEN (1975). The observed smooth response functions are related to the field components by

$$\begin{pmatrix} E_A \\ E_B \end{pmatrix} = \begin{pmatrix} Z_{AD} & Z_{AH} \\ Z_{BD} & Z_{BH} \end{pmatrix} \begin{pmatrix} B_D \\ B_H \end{pmatrix}$$

and the response functions are tabulated in Table 4 by equal increments of the square root of the frequency where E_A and B_D are pointed toward magnetic east and E_B and B_H are pointed toward magnetic north. Magnetic north is $-15°$ west of north, geographic.

For these data we assume that the island locally perturbs the electric field. The model used to interpret the data is given by the following:

$$\begin{pmatrix} E_A \\ E_B \end{pmatrix} = Z(f)D \begin{pmatrix} A+C & -1+B \\ 1+B & A-C \end{pmatrix} \begin{pmatrix} B_D \\ B_H \end{pmatrix}.$$

Table 4. The response tensor at Bermuda (in m/s).

f (cph)	Z_{AD}		Z_{AH}		Z_{BD}		Z_{BH}	
	Real	Imag	Real	Imag	Real	Imag	Real	Imag
1.00	−25.5	27.6	−21.4	35.4	30.0	−30.1	19.2	−29.6
1.25	−25.1	32.2	−29.6	43.2	34.3	−34.6	20.7	−29.7
1.52	−25.8	36.1	−37.4	49.5	39.2	−39.5	22.6	−30.3
1.83	−27.4	39.3	−44.4	54.5	44.2	−44.6	25.0	−31.0
2.16	−30.0	42.0	−50.8	58.3	49.0	−50.0	27.8	−31.8
2.52	−33.2	44.5	−56.7	61.1	53.6	−55.5	31.0	−32.5
2.91	−36.9	47.0	−62.0	63.0	57.7	−60.9	34.6	−32.8
3.32	−40.7	49.6	−66.9	64.2	61.2	−66.1	38.3	−33.0
3.76	−44.4	52.7	−71.2	64.9	64.1	−71.1	42.0	−32.9
4.24	−47.7	56.4	−75.1	65.5	66.3	−75.8	45.5	−32.8
4.73	−50.4	60.9	−78.6	66.1	67.9	−80.4	48.5	−32.8
5.26	−52.2	66.3	−81.8	67.2	69.0	−85.0	50.9	−33.4
5.81	−53.2	72.3	−85.1	68.9	69.8	−89.8	52.7	−34.8
6.39	−53.7	78.6	−88.8	71.1	70.4	−95.2	54.1	−37.2
7.00	−54.2	84.1	−93.5	73.4	71.2	−101.4	55.5	−40.4

The expression is analogous to Eq. (5.4). The response function $Z(f)$ is assumed to be that of an isotropic, horizontally layered mantle. The multiplier D has been explicitly introduced to indicate that the impedance is expected to be amplified by the presence of a nonconducting island. We expect the distortion and amplification parameters A, B, C, and D to be real and frequency independent (LARSEN, 1975). The parameters have been estimated as follows: for a horizontally layered mantle:

$$Z_D = ZD = (Z_{BD} - Z_{AH})/2$$

$$2A = \sum (Z_{AD} + Z_{BH}) Z_D^* / \sum |Z_D|^2$$

$$2B = \sum (Z_{AH} + Z_{BD}) Z_D^* / \sum |Z_D|^2$$

$$2C = \sum (Z_{AD} - Z_{BH}) Z_D^* / \sum |Z_D|^2$$

the sums being taken over all frequencies.

By applying the inequality constraints (WEIDELT, 1972) on Z, it is found that only for the frequency range 1.0 cpd$<f<$7.0 cpd is it possible to interpret Z in terms of a horizontally layered conductivity structure. The values of Z are given in Table 5 and the model misfit is based on the root mean square of the difference between the observed Z_{AD}, Z_{AH}, Z_{BD}, and Z_{BH} and the predicted model values using the estimates of the surface parameters A, B, and C and mantle response Z. The model misfit in Z is based on assuming A, B, C are known and the overall misfit is about 4%. On the other hand, if we assume Z is known perfectly, then the model misfit in A, B, and C is 1%.

The logarithmic response is defined by

$$U = 2 \log [(i\mu\sigma_0/\omega)^{1/2} Z]$$

with $\sigma_0 = 1$ mho/m. The model misfit in U is determined from the misfit in Z. The conductivity profile that fits U is given in Fig. 12 and the inversion misfit in U is found

Table 5. Mantle response function Z and logarithmic function U.

$A=-0.09$ $B=-0.01$ $C=-0.66$ $D=1.5$

f (cpd)	Z(m/s) Real	Z(m/s) Imag	Z(m/s) Model misfit	U Real	U Imag	U Model misfit
1.00	18.9	−23.9	1.9	2.77	−0.24	0.13
1.25	21.9	−27.6	1.4	2.84	−0.23	0.08
1.52	25.2	−30.9	1.2	2.89	−0.20	0.06
1.83	28.6	−33.9	1.2	2.92	−0.17	0.05
2.16	32.0	−36.6	1.2	2.94	−0.13	0.05
2.52	35.4	−39.1	1.1	2.95	−0.10	0.04
2.91	38.7	−41.2	1.1	2.95	−0.06	0.04
3.32	41.9	−43.2	1.3	2.93	−0.03	0.04
3.76	44.7	−45.0	1.6	2.92	−0.01	0.05
4.24	47.1	−46.8	2.1	2.89	−0.01	0.06
4.73	49.1	−48.8	2.6	2.86	−0.01	0.08
5.26	50.8	−51.1	3.2	2.84	−0.01	0.09
5.81	52.1	−53.9	3.7	2.82	−0.03	0.10
6.39	53.4	−57.0	4.3	2.80	−0.06	0.11
7.00	55.0	−60.4	4.8	2.80	−0.09	0.12

to be 0.02. Here D has been set to 1.5 to bring the depths of layers in the conductivity profile into harmony with the depths of layers derived from the Station 20 analysis (Fig. 12). A variation in D shifts the profile parallel to the D shift line in the figure. For an island consisting of a nonconducting, circular cylinder of rock, the appropriate value would be $D=2.0$, while for an island equally conductive as sea water D would be 1.0. The adopted value, lying between these extremes, is entirely reasonable.

8. Discussion

Electrical conductivity models constructed from the impedance measurements at the two sea floor sites and on Bermuda are remarkably similar. The oceanic impedance measurements indicate a highly conducting and anisotropic upper layer below the ocean. The hard direction of conductance is roughly east-west and is consistent with the known properties of sediments (conductivity 1 S/m, thickness 2 km in the Hatteras abyssal plain) and water saturated basalt lava flows (conductivity less than 0.1 S/m, 1.5 km thickness, Kirkpatrick, 1979). On the other hand the easy direction (north-south) is so highly conducting that it is difficult to conceive that the conductance lies entirely in these layers. These results are tentative because there has been as yet no examination of the scales of the ionospheric source fields as reflected in the spatial coherence of the measured fields, nor of the large scale structure of the ocean. Both scale effects could have different influences on the north-south and east-west impedances at the sea bed. The validity of the one dimensional models depends on a certain degree of conductive contact between the ocean and the basement conductive layers, since otherwise the telluric current flowing in the ocean will be trapped by poorly conducting continental boundaries. In our models we have assumed that the flow is unimpeded laterally.

The sea bed impedance measurements are supplemented by the Bermuda analysis for examination of the deeper conductive structures. (The island impedance is insensitive to the sea bed crustal rocks because of the thick oceanic shield surrounding Bermuda.) All analyses are consistent with sharp rises in conductivity at 75 and 220 km. The analysis of the vertical gradient of the horizontal magnetic fluctuations by POEHLS and VON HERZEN (1976) is also consistent although their model has less detail. The agreement is encouraging. It seems to us that discovering the lateral extent of these structures will contribute to understanding of the nature and form of the lithosphere-asthenosphere transition.

We have demonstrated the electrical signal of water movements in the deep sea. The low frequency "mesoscale eddy" motion unfortunately provides EMF's which override the ionospheric signal. This fact will make deep impedance soundings difficult. It is known, however, that the water movements are strongest within the westward edges of the North Atlantic and Pacific. It is likely therefore that impedance probing of deeper structures under the sea will be most successful away from these noisy regions.

The electric field from these very low frequency water movements has provided a rough upper limit to the conductivity just below the ocean. This value is slightly inconsistent with the north-south conductivity inferred from the impedance measurements. The discrepancy is unimportant because the conductance of rocks in the vertical, which determines the water current effect may well be lower than the horizontal conductance.

This research was supported by grants from the US NSF and ONR. R. W. was supported by grants from the German Academic Exchange Program. K.A.P. was supported by NASA Grant NSG 7002 during much of the analysis. C. S. C. is grateful for the hospitality of the Institut fur Geophysik, Goettingen during a critical time while the analysis was carried out. Discussion with U. Schmucker and P. Weidelt was particularly helpful.

REFERENCES

COX, C. S., J. H. FILLOUX, and J. C. LARSON, Electromagnetic studies of ocean currents and the electrical conductivity below the seafloor, in *The Sea*, Vol. 4, edited by Maxwell, Ch. 17, pp. 637–693, John Wiley and Sons, New York, 1970.

COX, C. S., J. H. FILLOUX, D. CAYAN, and P. PISTEK, Electromagnetic observations during Mode: Sea floor electric field and transport, unpubl. ms., 1978.

FILLOUX, J. H., Electric field recording on the sea floor with short span instruments, *J. Geomag. Geoelectr.*, **26**, 269–279, 1974.

FREELAND, H. J. and W. J. GOULD, Objective analysis of meso-scale ocean circulation features, *Deep-Sea Res.*, **23**, 915–923, 1976.

KIRKPATRICK, R. J., The physical state of the oceanic crust: Results of downhole geophysical logging in the Mid-Atlantic Ridge at 23°N, *J. Geophys. Res.*, **84**, 178–188, 1979.

LARSEN, J. C., Low frequency (0.1–6.0 cpd) electromagnetic study of deep electrical conductivity beneath the Hawaiian Islands, *Geophys. J. R. Astr. Soc.*, **43**, 17–46, 1975.

POEHLS, K. A. and R. P. VON HERZEN, Electrical resistivity structure beneath the Atlantic Ocean, *Geophys. J. R. Astr. Soc.*, **47**, 331–346, 1976.

WINTER, R., Analysis and first interpretation of magnetotelluric soundings on the deep sea floor of the North Atlantic, 3rd workshop on electromagnetic induction in the earth, Abstr., Sopron, Hungary, 1976.

WEIDELT, P., The inverse problem of geomagnetic induction, *Z. Geophys.*, **38**, 257–289, 1972.

North Pacific Magnetotelluric Experiments

J. H. FILLOUX

Scripps Institution of Oceanography, University of California, San Diego, La Jolla, California, U.S.A.

(Received December 12, 1977; Revised March 31, 1978)

Electromagnetic signals from two deep sea floor magnetotelluric stations located far from coastal influence in the North Central and North Eastern Pacific are evaluated.

The electric field was recorded by means of salt bridge chopper type instruments with a resolution of .01 μV/m. Independent magnetic variographs using suspended magnet sensors recorded magnetic variations accurate to .2 γ. Both electric and magnetic recorders are characterized by very small long term drift. They are self contained, store data on magnetic cassettes and travel free, between surface and bottom.

Magnetotelluric interpretation of records from a first station, located 450 nautical miles to the NNE of Hawaii, at a position which corresponds to magnetic reversal anomaly 31 and to a plate age of 72 m.y. suggests a moderate to low lithospheric conductivity with an average value less than 3×10^{-3} (ohm m)$^{-1}$ over the upper 60 km, increasing sharply around 100 km. A high conductivity tongue, with an average value around 6×10^{-2} (ohm m)$^{-1}$ occurs over the interval 180 ± 40 km. Meaningful information ceases for a depth beyond 400 km.

A second and more recent magnetotelluric sounding 700 km off central California suggests that a very high and quite shallow conductivity feature implied from earlier work was probably overemphasized. Nevertheless, according to the new data a relatively higher conductivity at a shallower depth seems to differentiate this second station from the first. This result is in harmony with the younger crustal age at the second station (32 m.y.; magnetic reversal 12–13).

1. Introduction

Electromagnetic investigation of the earth by interpretation of relationships between natural electric and magnetic pulsations observed at its surface has come a long way since SHUSTER (1889) demonstrated the necessarily high electrical conductivity of its interior. While magnetotelluric soundings of crust and upper mantle have been performed on land almost all over the world following TIKHONOV (1950) and CAGNIARD (1953), electromagnetic exploration of the sea floor has only begun.

Magnetotelluric (MT) soundings of the oceanic basement presently available are restricted to the northern hemisphere, with one experiment in the Atlantic and a few in the eastern and central Pacific.

A limited MT experiment was carried out in 1965, 630 km offshore from central California at the extreme point of an electromagnetic traverse across this coastal area (SCHMUCKER, 1964; FILLOUX, 1967a; COX *et al.*, 1971). The position of the station, at that time labeled S. F. (for sea floor) is shown on the map of Fig. 1. Electric field measure-

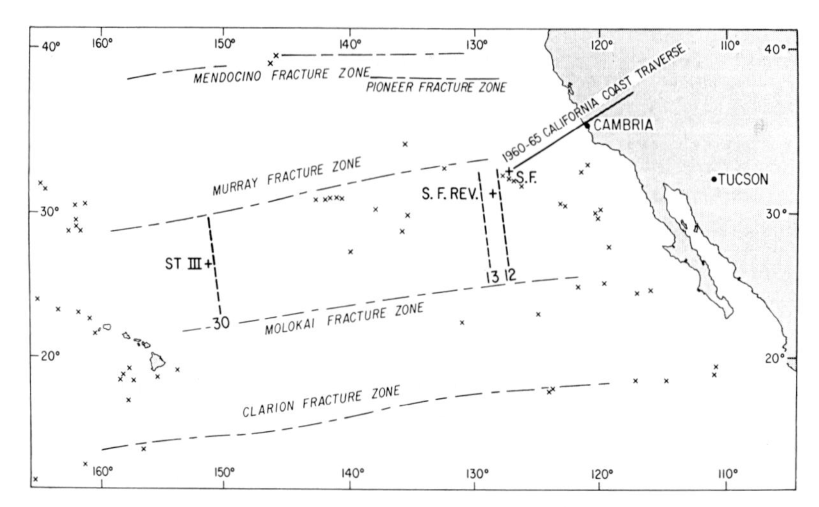

Fig. 1. Location of the various stations. Farewell to Aggy, St. III, 450 nautical miles to the NNE of Hawaii; S. F., our 1965 station off California at the seaward end of an electromagnetic traverse across the coastal area; S. F. Rev. near S. F. but displaced away from the Sieberling line of seamounts. The significant magnetic reversal anomalies are also indicated as well as the corresponding crustal age (after ATWATER and MENARD, 1970). The crosses indicate prominent sea mounts.

ments were then carried out by means of long cables stretched on the sea floor while a single horizontal component of magnetic variations, to the magnetic east, was recorded at the same place, though not simultaneously (FILLOUX, 1967a, b). The relationships between the non-simultaneous electric and magnetic signals were derived by reference to two other stations. The resulting interpretation suggested an exceptionally high upper mantle conductivity occurring at an extremely shallow depth, namely .4 (ohm m)$^{-1}$ at 30 km (FILLOUX, 1967a).

Although performed on islands rather than on the sea floor, Larsen's experiments lead to a deep MT sounding of the oceanic mantle below and adjacent to the Hawaiian Island chain (LARSEN, 1975). During the MODE-I (Mid Ocean Dynamic Experiment), two MT soundings resulted from a multi-institutional study of mid oceanic circulation in the Atlantic (COX *et al.*, 1980).

More recently two new MT experiments have been performed in the Pacific using clusters of novel instruments which permitted certain improvements in performance and a good degree of redundancy. Although still far from optimum, these experiments provide a solid ground upon which to assess the true merits of ocean floor MT techniques. The object of this report is to review the significance of these experiments at the present stage of data analysis and interpretation.

2. Instrumentation

The experimental equipment involved individual, short span, two horizontal component electric field recorders with electrode drift removal by electrode switching and indepedent, three component magnetic variographs. Electrode limitations as well as the various aspects of electrode construction, electrode noise and drift, as well as drift rejection by means of salt bridge choppers have been reviewed in some detail elsewhere (FILLOUX, 1967a, 1973, 1974). The difficult problems associated with operation on the sea floor of salt bridge choppers have been described more recently (FILLOUX, 1980). Performance of present electric field recorders result in a least count of 0.01 V/m^{-1} for 6 m separation between sampling points. The magnetic variographs are of the classic suspended magnet type; however, they rely upon modern and highly sensitive optometric techniques to resolve the extremely tenuous rotations resulting from very stiff- and sturdy-suspension fibers (FILLOUX, 1980).

3. First Experiment

The first set of data referred to above was collected in the fall of 1976 at a position 450 nautical miles to the north/northeast of Hawaii (see map Fig. 1: Station III, cruise "Farewell to Aggy", 26°32'N, 151°20'W). Of eight instruments implanted at this location, seven were recovered providing two redundant sets of electric and magnetic data. A typical example of the signals recorded is shown on Fig. 2, for a duration of one week (for early report, see FILLOUX, 1977).

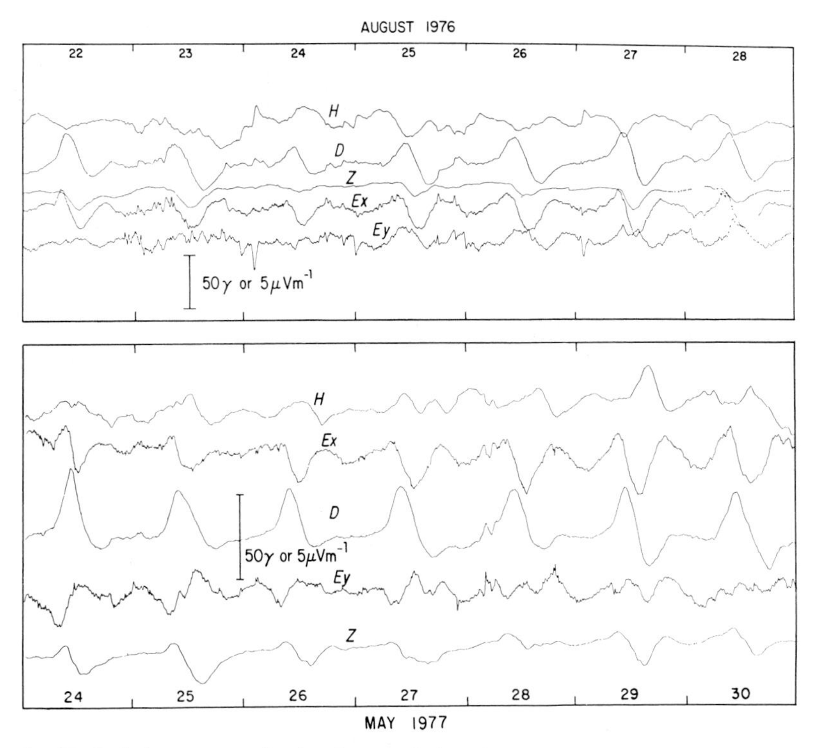

Fig. 2. Typical electromagnetic signals recorded at the sea floor at St. III, upper box and at S. F. Rev., lower box. Notice that the vertical scales are different in the two cases.

The spectra of horizontal electric and magnetic fluctuations for a three week simultaneous coverage are shown on Fig. 3. Since substantial contributions to the signals result from oceanic motions at tidal frequencies, the spectral estimates have been distributed in a pattern that eliminate tidal bands. Furthermore a fewer number of Fourier components are averaged, in the lower frequency bands, to better preserve the possible curvature of the impedance functions.

Quite noticeable is the steep slope of all spectra, but particularly of the magnetic fluctuations toward high frequencies, largely a consequence of oceanic shielding by 5.5 km of sea water. This unavoidable effect constitutes a strong limitation to the study by MT methods of the oceanic crustal layers. Also to be noticed is the excess energy in the semidiurnal tidal band, conspicuous in both the electric and magnetic spectra.

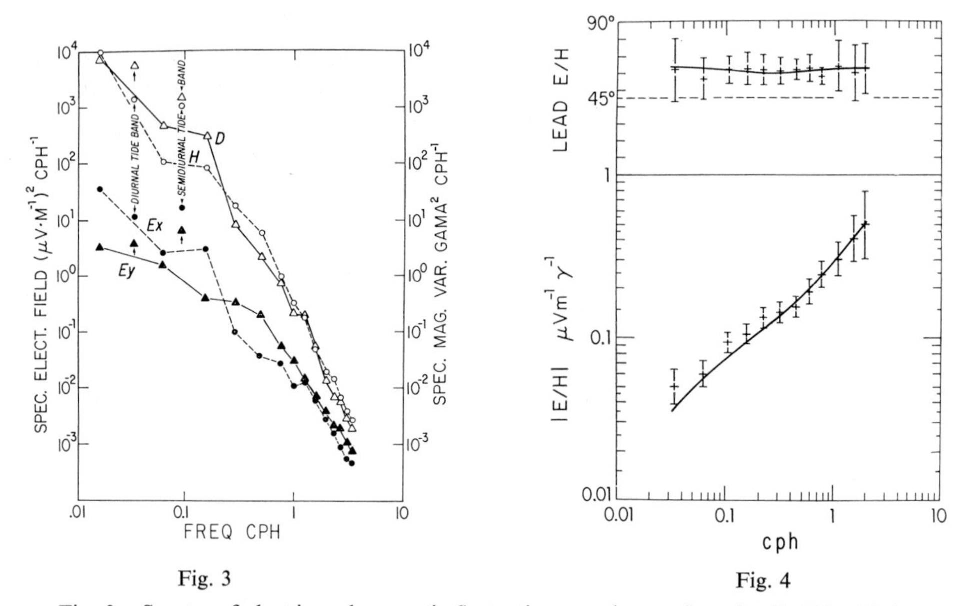

Fig. 3. Spectra of electric and magnetic fluctuations on the sea floor for St. III. Notice the extra energy in the semidiurnal tidal band and the steep slope of the spectra toward high frequencies, particularly for the magnetic variations.

Fig. 4. Impedances of sea floor derived from observations at St. III, in modulus, lower box and in phase, upper box. The range of validity derived from coherence and number of degrees of freedom in each band is shown by error bars (95% confidence limit).

The transfer functions between horizontal electric and magnetic pulsations (or impedance of the sea floor) are shown on Fig. 4, together with their uncertainty at the 95% confidence level. With B representing the horizontal magnetic fluctuation vector (H, D) and E representing the horizontal electric vector (Ex, Ey), the impedance Z and admittance A of the sea floor are defined by

$$E(f)=Z(f)B(f) \tag{1}$$

$$B(f)=A(f)E(f)\,, \tag{2}$$

where (f) indicates frequency dependance.

Because the estimated impedances and admittances may depend upon azimuth, a meaningful one dimensional interpretation of conductivity structure with depth could be justified only if the observed anisotropy is small. This condition is reasonably well satisfied here since the ratios of impedance maxima to minima remain relatively moderate as shown on Table 1 for four representative frequencies. The retained impedances Z'

Table 1. Polarization of impedance tensor for four frequencies. L/l ratio of maxima to minima. Az orientation of major axii with respect to geographic north, e.g. 90° for Eastward (Station III, Farewell to Aggy).

Frequency (cph)	L/l	Az (degree)
.0615	1.44	104
.3193	1.37	128
.5967	1.21	94
.9883	1.48	106

are taken to combine the extreme values of Z and A according

$$Z'=(Z_{xD}-Z_{yH})^{1/2}(A_{Hy}-A_{Dx})^{-1/2}$$

where Z_{xD}, Z_{yH} and A_{Hy}, A_{Dx} are the off-diagonal elements of Z and A. This choice is related to a similar treatment by BERDICHEVSKIY and DMITRIEV (1976) involving only the impedances Z. These authors have shown that it tends to minimize the effect on the impedance estimates of distorsion in the electric current pattern associated with simple and small superficial heterogenieties. It also assumes equipartition of noise into electric and magnetic signals, an assumption probably not optimum: our understanding of oceanic processes suggests that motional fields associated with displacement of sea water across the earth's magnetic field results in electromagnetic signals contaminating far more adversely the electric than the magnetic records. Data interpretation tends to confirm this situation indirectly: conductivity models optimizing the data best tend to require that the lower frequency impedances be consistently lower than the observed ones (see Fig. 4).

An interpretation of the impedance estimates of Fig. 4 in terms of a 5 layer horizontally stratified and isotropic conductivity model is shown on Fig. 5 together with a proposed range of validity. The model fitting technique has been outlined briefly earlier (FILLOUX, 1977) and a more thorough description and discussion is in preparation. Basically the impedances of a model with arbitrarily selected thicknesses are calculated and compared to the observed impedances. The initially ascribed conductivities then are modified in a way to improve the fit till an optimum, based on the behavior of the variance between observations and predictions, is reached. The range of allowable conductivity in each layer is based on the sensitivity of the variance to conductivity changes.

In order to check the stability of the modeling method, models with a variety of numbers of layers, a variety of layer thicknesses have been considered using a vast range of initial conductivity values. For illustration an optimum 12 layer model is also shown in Fig. 5. The advantage of considering models with an otherwide unjustifiably high number of thinner layers is that the most conspicuous features of the optimum conductivity pro-

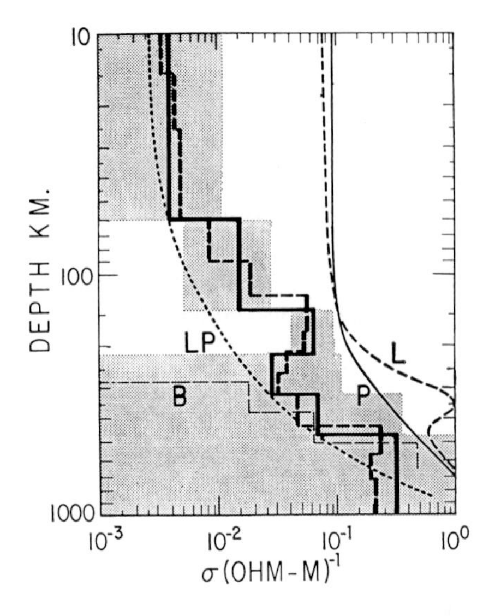

Fig. 5. Proposed optimum models of conductivity structure at St. III. (1) 5 layers model plus constant conductivity substratum, heavy continuous line, with range of validity, grey area, and (2) 12 layer model, heavy dashed line. Conductivity dependences with depth obtained by other researchers are also shown: LAHIRI and PRICE (1939), LP; PARKER (1971), P; BANKS (1972), B; LARSEN (1975), L.

file can be used to select more readily a layering geometry which preserves best these features. Further modification of the layer thicknesses can be suggested by the conductivity error bars, the selection of thinner layers resulting in increased uncertainty of the estimated average conductivity in these layers.

For reference, conductivity structures proposed by other researchers are also shown on Fig. 5, namely: LAHIRI and PRICE (1939), (LP); BANKS (1972), (B); PARKER (1971), (P); and LARSEN (1975), (L).

Both 5 and 12 layers models lead to similar interpretation. The upper lithospheric layer appears to be of low to moderate conductivity at the highest. This is a poorly defined area, with broad uncertainty. Below, the conductivity increases, first slowly, then beyond 60 km more rapidly with depth.

The best conductivity resolution ranges over 50 to 250 km depth. The most conspicuous and significant feature appears to be the sharp rate of conductivity increase occurring around 110–140 km, with high conductivity values sustained below in a layer 80 to 100 km thick. A temporary slope reversal, not fully supported within our error bar criterion, possibly follows.

It should be noted that the conductivity estimates are meaned to represent the averaged conductivity in each layer and that the occurrence of thin strata with much higher conductivity is not excluded as long as their thickness remains correspondingly small.

Geophysical interpretation should be pursued with caution. (1) Electrical conductivity depends on several parameters: material composition, heterogeneity, particularly that related to liquid inclusions—water or magmas—, and temperature. (2) The most descriptive techniques of geophysical investigation of crust and upper mantle, namely seismology, tends to weaken in the area where the conductivity structure is best resolved, as lower velocity, higher absorption layers are encountered. (3) The scarcity of similar MT data rules out the possibility at this time of much comparative work.

Our conductivity estimates are intermediate between those of Lahiri and Price (LP) almost strictly dependent on continental data, and those of Larsen (L), derived from observations in the Hawaiian Islands and possibly representative of the deep Pacific basement adjacent to this island chain. They are also intermediate between those of Banks (B) and Parker (P), although the meaning of this relationship is not clear since the B and P profiles appear to reflect differences in interpretation of similar data (PARKER, 1970; BANKS, 1972).

If one attempts to infer from our proposed conductivity profile the depth of the zone of maximum shearing strain in the asthenosphere over which the Pacific plate slides, one may be tempted to select the high conductivity zone centered between 140 and 220 km depth. No prominent feature other than a sharp conductivity increase around 120 km suggests a possible clue to the depth of the lithosphere-asthenosphere interface. This location appears, however, somewhat deeper than that established by other geophysical investigations. At the observation site the nearest reversal magnetic anomaly corresponds to number 31 (ATWATER and MENARD, 1970; HEIRTZLER *et al.*, 1968) with an implied age of 72 m.y. (BERGGREN, 1972). For such a crustal age, LEEDS *et al.* (1974) predict a lithospheric thickness of about 70 km. Similar values are mentioned in other studies (PRESS, 1970; LEEDS, 1975) or are consistent with other research (CHAPMAN and POLLACK, 1977; YOSHII, 1975). The divergence may simply result from our present inability to provide an unambiguous definition of the asthenosphere and to identify its boundaries in terms consistent with the depth dependence of the various parameters accessible to geophysical investigation. We note in PRESS (1970) the somewhat arbitary definition of the lithosphere-asthenosphere interface as the depth at which the shear velocity drops below 4.5 km/s. This depth, very dependent on the velocity criterion, is around 70 km. However, in the same paper, the rather broad shear velocity minimum, around 4.2 km/s which spans the range 130–260 km (for oceanic data) matches rather well the conductivity maximum of our proposed conductivity structure (140–220 km). A satisfactory resolution of these uncertainties shall probably have to wait for further cooperative work in all aspects of submarine geology and in laboratory research on rocks at high pressure and temperature.

It is interesting to note that both Larsen's profile (L) for the Hawaii area and ours (Fig. 5), are characterized by a high conductivity maximum. Since it is not possible to associate the one in the Larsen profile to the asthenosphere on account of its excessive depth it is far from obvious that these two features could be closely related. LARSEN (1975) has shown that a scaling constant in the estimate of the telluric distortion pattern caused by the island of Oahu where his observations were made remains uncertain. However no change in this scaling constant could bring the two conductivity profiles closer: the required reduction of the depth scale in the L model would result in grossly enlarging the conductivity scale, thus increasing further the already high conductivity contrast—a factor of 10—between his and our maxima. Perhaps the deep and very high conductivity layer in Larsen's model relates to some yet unknown process associated with the existence of the island chain, or results from a more complex distortion of the telluric current pattern than assumed (due for instance to important vertical current flow associated with island heterogeneity).

4. Second Experiment

A second experiment was carried out more recently in May–July 1977 for the primary purpose of engineering tests but also for checking on a previous magnetotelluric sounding at Station S. F., off central California, performed in 1965; see map on Fig. 1; (FILLOUX, 1967a; COX *et al.*, 1971). The merit of this early experiment was that, no matter how imperfect, it constituted a hopeful step toward sea floor magnetotelluric exploration of the oceanic basement. The equipment used then was quite archaic compared to that available today. It involved two long lines (1 km) stretched onto the sea floor and a single component magnetic variograph recording the horizontal magnetic variation to the magnetic east only (FILLOUX, 1967b). Additional weaknesses resulted from the non-simultaneity of electric and magnetic records and the exceptionally low level of ionospheric activity

The non-simultaneous electric and magnetic records were cross correlated through the use of two intermediate magnetic variation records spanning all experiments, one from the temporary station of Cambria, the other from the Tucson observatory. This early magnetotelluric sounding predicted an unexpectedly high conductivity, around .4 (ohm/m)$^{-1}$ at a depth as shallow as 30 km. Some inconsistencies, however, could not be re-

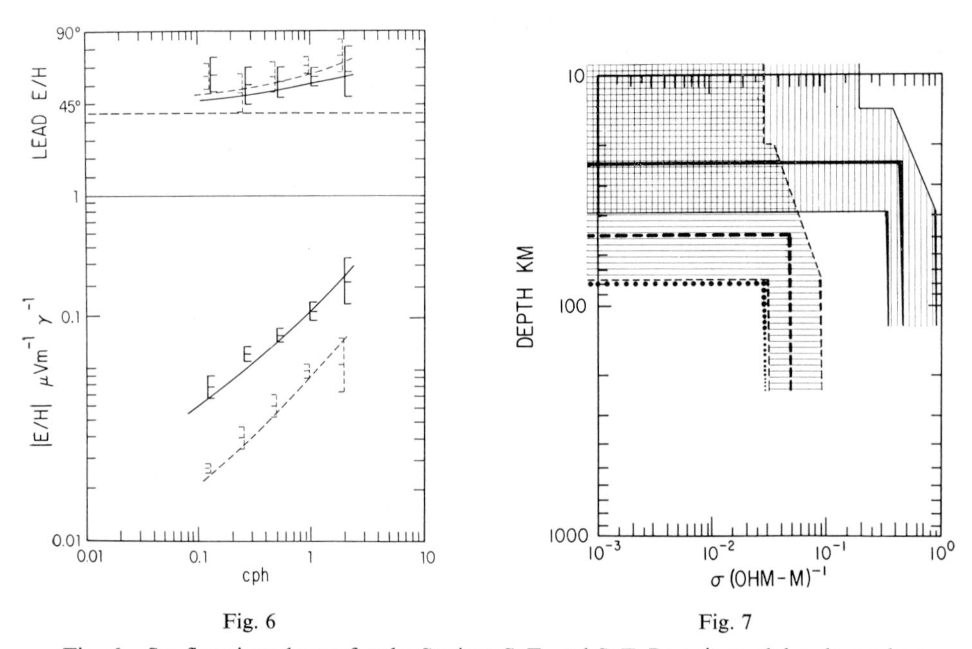

Fig. 6 Fig. 7

Fig. 6. Sea floor impedances for the Stations S. F. and S. F. Rev., in modulus, lower box and phase upper box. Dashed line and left tics for error bars in S. F. estimates. Continuous line and right tics for S. F. Rev. estimates.

Fig. 7. Two layer model interpretation of impedances of sea floor at Station S. F. (continuous heavy line for optimum model, vertical stripes for uncertainty limits) and at Station S. F. Rev. (heavy dashed line for optimum model, horizontal stripes for uncertainty limits). The optimum two layer interpretation of sea floor impedances at Station III is also shown (circular dots) for comparison.

solved (FILLOUX, 1967a) and further questioning of the results arose (GREENHOUSE, 1972; LAUNAY, 1974). All aspects of these difficulties suggested an underestimation of the sea floor impedance, and, in turn an underestimation of the electric field. The latter could have resulted from an inadequate stretching of the cables, a possibility that had been dismissed (for discussion, see FILLOUX, 1967a).

For the reasons mentioned above we found it desirable that these early results, often used and quoted, be confirmed or corrected. A necessity to test and to evaluate modified instrumentation in deep water close to California provided an opportunity.

Because of the proximity of a chain of sea mounts near the Station S. F. the new selected site—S. F. Revisited (see map of Fig. 1) was displaced slightly farther offshore. Three pairs of electric and magnetic recorders were used, resulting in a high degree of confidence in the data.

Because of the recent recovery of the instrumentation and because of the necessity to handle a large volume of data, the analysis has only begun. Since a preliminary report may be useful, and to make the comparison of the early and of the new data more readily meaningful, we have treated a segment of the data from S. F. Revisited in a manner identical to the earlier treatment of the S. F. data, in fact using the very computer programs produced for this purpose. In accordance with this approach, we have not used the formerly unavailable northward magnetic variation component (for detail, see FILLOUX, 1967a). The interpretation involves fitting the impedance of a three parameter, layered and isotropic model to the observed sea floor impedance. The model parameters represents the thickness d of a uniformly conducting upper layer on top of an infinitely deep second layer, and the conductivities σ_{up} and σ_{down} of the two layers.

The observed impedances for both S. F. and S. F. Revisited are shown on Fig. 6, with their 95% confidence limits. The evidence of an underestimate of the impedances in the S. F. experiment is overwhelming. Since the scale of the magnetic data could be closely checked by comparing the daily variation signals on the sea floor with those on the adjacent land, and since this scale is further confirmed by the S. F. Revisited data, underestimation of the electric field in the S. F. experiment appears inescapable. Comparison of the new and old data suggests a factor of 2 or slightly more. In the absence of a better explanation the possibilities of an inadequate stretching of the electric cables or of an excessive loading of the silver-silver chloride electrodes, both previously rejected, must be reconsidered.

Interpretation of the impedances in terms of the three parameter model described earlier is illustrated by the diagrams on Fig. 7. As expected, the step increase of conductivity with depth is less sharp and occurs deeper in the S. F. Revisited optimum model than in the earlier S. F. profile (.05 $(\mathrm{ohm/m})^{-1}$ at 50 km against .4 $(\mathrm{ohm/m})^{-1}$ at 30 km). The relative uncertainty is roughly the same in both cases.

5. Conclusion

The two layer interpretation of the data from the first experiment (Station III, Farewell to Aggy) is also shown on Fig. 7, with $\sigma_{\mathrm{down}}= .03(\mathrm{ohm/m})^{-1}$, $\sigma_{\mathrm{up}} \ll \sigma_{\mathrm{down}}$ and $d=$ 80 km (uncertainty on conductivity not shown, but typical of other two cases).

The range of validity of both S. F. Revisited and Station III interpretations have a substantial overlap. This, however, is largely due to the coarseness of a two layer modeling of sea floor structures. If so, the generally higher conductivity at a shallower depth encountered at S. F. Revisited is quite significant. This is an encouraging result, consistent with the younger age of the Pacific plate at S. F. Rev., namely 32 m.y. (magnetic reversal 12–13), against 72 m.y. for Station III. A more thorough treatment of the S. F. Rev. data now in progress should lead to a more specific intercomparison of the two sea floor structures.

This work was supported by the U.S.-N. S. F., Grants OCE76-81867 and OCE77-20488.

REFERENCES

ATWATER, T. and H. W. MENARD, Magnetic lineation in the Northeast Pacific, *Earth Planet. Sci. Lett.*, **7**, 445–450, 1970.

BANKS, R. J., The overall conductivity distribution of the earth, *J. Geomag. Geoelectr.*, **24**, 337–351, 1972.

BERDICHEVSKIY, M. N. and V. I. DMITRIEV, Distortion of magnetic and electrical fields by near surface lateral inhomogeneities, in *Geoelectric and Geothermal Studies*, KARP Geophysical Monograph, Acta Geodaet., Geophys. et Montanist Acad. Sci. Hung., 11, (3–4), 447–483, 1976.

BERGGREN, W., A Cenozoic time scale-some implications for regional geology and paleobiogeography, *Lethaia*, **5**, 195–215, 1972.

CAGNIARD, L., Principi de la methode magentotellurique, *Ann. Geophys.*, **9**, 95–125, 1953.

CHAPMAN, D. S. and H. N. POLLACK, Regional geotherms and lithospheric thickness, *Geology*, **5**, 265–268, 1977.

COX, C. S., J. H. FILLOUX, and J. C. LARSEN, Electromagnetic studies of ocean currents and electrical conductivity below the ocean floor, in *The Sea*, edited by A. E. Maxwell, Vol. 4, pp. 637–693, Wiley-Interscience, New York, 1971.

COX, C. S., J. H. FILLOUX, D. I. GOUGH, J. C. LARSEN, K. A. POEHLS, R. P. VON HERZEN, and R. WINTER, Atlantic lithospheric sounding, *J. Geomag. Geoelectr.*, 1980 (this issue).

FILLOUX, J. H., Oceanic electric currents, geomagnetic variations and the deep electrical conductivity structure of the ocean continent transition of Central California, Ph. D. Thesis, University of California, San Diego, 1967a.

FILLOUX, J. H., An ocean bottom, *D* component magnetometer, *Geophysics*, **32**, 978–987, 1967b.

FILLOUX, J. H., Techniques and instrumentation for study of natural electromagnetic induction in the sea, *Phys. Earth Planet. Inter.*, **7**, 323–338, 1973.

FILLOUX, J. H., Electric field recording on the sea floor with short span instruments, *J. Geomag. Geoelectr.*, **26**, 269–279, 1974.

FILLOUX, J. H., Ocean floor magnetotelluric sounding over North Central Pacific, *Nature*, **269**, 297–801, 1977.

FILLOUX, J. H., Observation of very low frequency electromagnetic signals in the ocean, *J. Geomag. Geoelectr.*, 1980 (this issue).

GREENHOUSE, J., Geomagnetic time variations on the sea floor off Southern California, Thesis (SIO Ref. 72–67), Univ. of California, San Diego, 1972.

HEIRTZLER, J. R., G. O. DICKSON, E. M. HERRON, W. C. PITMAN III, and X. LE PICHON, Marine magnetic anomalies, geomagnetic field reversals and motions of ocean floor and continents, *J. Geophys. Res.*, **78**, 7752–7762, 1968.

LAHIRI, B. N. and E. T. PRICE, Electromagnetic induction in nonuniform conductivity of the earth from terrestrial magnetic variations, *Phil. Trans. Roy. Soc. Lond.*, **A237**, 509–540, 1939.

LARSEN, J. C., Low frequency (0.1–6. cpd) electromagnetic study of deep mantle electrical conductivity benath the Hawaiian Islands, *Geophys. J. R. Astr. Soc.*, **43**, 17–46, 1975.

LAUNAY, L., Conductivity under the oceans: interpretation of a magnetotelluric sounding 630 km off the California coast, *Phys. Earth Plalnet. Inter.*, **8**, 83–86, 1974.

LEEDS, A. R., Lithospheric thickness in the western Pacific, *Phys. Earth Planet. Inter.*, **11**, 61–64, 1975.
LEEDS, A. R., L. KNOPOLF, and E. G. KAUSEL, Variations of upper mantle structure under the Pacific Ocean, *Science*, **18**, 141–143, 1974.
PARKER, R. L., The inverse problem of electrical conductivity in the mantle, *Geophys. J. R. Astr. Soc.*, **22**, 121–138, 1970.
PRESS, F., Regionalized earth models, *J. Geophys. Res.*, **75**, 6675–6681, 1970.
SCHMUCKER, U., Anomalies of geomagnetic variations in the southwestern United States, *J. Geomag. Geoelectr.*, **15**, 193–221, 1964.
SHUSTER, A., The diurnal variations of terrestrial magnetism, *Phil. Trans. Roy. Soc.*, **A180**, 457–518, 1889.
TIKHONOV, A. N., Determination of the electrical characteristics of deep layers of the earth's crust, *Doklady Akad. Nauk.* (*SSSR*), **73**, 275, 1950.
YOSHII, R., Regionality of group velocities of Rayleigh waves in the Pacific and thickening of the plate, *Earth Planet. Sci. Lett.*, **25**, 305–312, 1975.

Note added in proof

The data from the Station S.F. Revisited have been analyzed in greater detail since this paper was written, see following reference:

FILLOUX, J. H., Magnetotelluric soundings over the Northeast Pacific may reveal spatial dependence of depth and conductance of the asthenosphere, *Earth Planet. Sci. Lett.*, **46**, 244–252, 1980.

Electromagnetic Induction in the Vancouver Island Region

W. NIENABER,* H. W. DOSSO,* L. K. LAW,** F. W. JONES,***
and V. RAMASWAMY***,†

**Department of Physics, University of Victoria, Victoria, B. C., Canada,*
***Division of Geomagnetism, Earth Physics Branch, Department of Energy, Mines and Resources, Victoria, B. C., Canada*
****Department of Physics and the Institute of Earth and Planetary Physics, University of Alberta, Edmonton, Alta., Canada*

(Received October 7, 1977; Revised February 12, 1978)

Electromagnetic induction in the Vancouver Island region of British Columbia, Canada, was studied using scaled analogue model measurements and field measurements from two pairs of stations spanning the channel between Vancouver Island and the continent. Good agreement between model measurements and field measurements for the vertical magnetic field component was obtained. The model results show that conductive telluric current-channeling in the shallow seawater channel is important for the frequencies studied. The results also show that the geometry of the channel affects the current channeling and that there is some diffusion of current into the continent where the ocean channel direction is not parallel to the electric field of the inducing field.

1. Introduction

A problem of interest in recent induction studies is the possibility of telluric currents flowing in conductive channels, such as ocean channels between islands and continents, but induced elsewhere. Evidence of such conductive channeling has been observed, for example, by COCHRANE and HYNDMAN (1974) in the St. Lawrence River region, by WHITHAM and ANDERSON (1965) in the Canadian Arctic, and by EDWARDS *et al.* (1971) in the shallow seas around the British Isles. On the basis of work done in the Vancouver Island area, it appears that the geomagnetic fields around Vancouver Island are also affected by electric currents induced in the deep ocean and flowing in the seawater channel around Vancouver Island.

In the present work, the behaviour of the electromagnetic fields in the Vancouver Island region is studied with the aid of laboratory analogue model measurements and measurements from four field stations. Vancouver Island is situated on a continental shelf within a short distance of the deep ocean. A shallow channel separates the island from the continent. For a 5 minute period the sea water channel is less than 1/20 of a skin depth deep while the deep ocean is approximately 1/3 of a skin depth deep. In the shallow ocean channel current channeling could be more important than local induction.

† Present address: Institute of Petroleum Exploration, Dehra Dun, India.

2. Measurements and Discussion of Results

The laboratory analogue model used in the present work, as well as the scaling conditions, have been described in detail previously (Dosso, 1966a, b, 1973) and will not be treated here. The length and frequency scaling factors are chosen so that 1 cm in the model represents 5 km, a model frequency of 30 kHz simulates a frequency of 0.004 Hz, and a model frequency of 3 kHz simulates 0.0004 Hz, in the geophysical system.

A scaled model of the Vancouver Island region was constructed having the shape shown in the simplified map of Fig. 1. The shallow ocean around the island and the nearby deep ocean were simulated by a graphite plate suitably shaped and machined to the

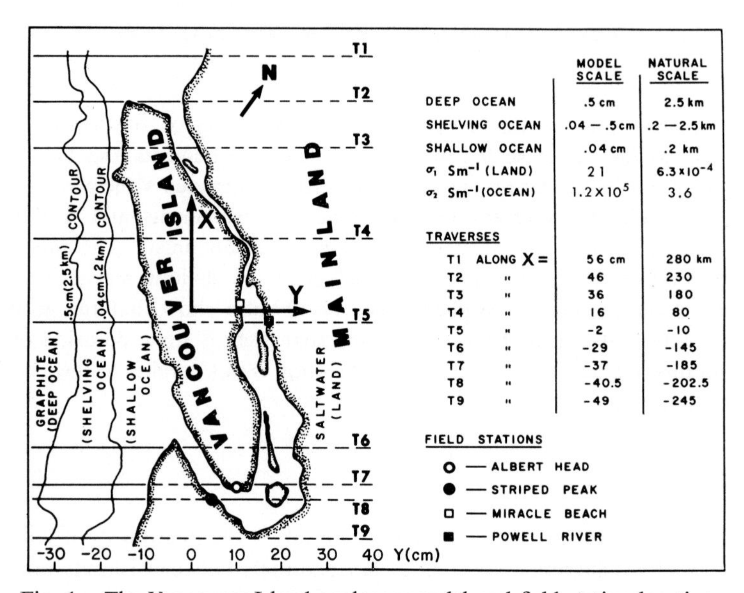

Fig. 1. The Vancouver Island analogue model and field station locations.

scaled island and the continental coastline contours and the various ocean depths. This graphite model was suspended at the surface of the salt solution in the model tank. The electric field of the overhead inducing uniform source was in the x-direction. Measurements of the amplitudes (E_x, E_y, H_x, H_y, H_z) and phases (ψ_x, ψ_y, ϕ_x, ϕ_y, ϕ_z) of the electric and magnetic field components for traverses over this Vancouver Island model were carried out. As well, field measurements were carried out for two pairs of stations, as shown in Fig. 1, spanning the ocean channel (Miracle Beach–Powell River, and Albert Head–Striped Peak) and the vertical magnetic field component at each station was obtained using the BAILEY *et al.* (1974) transfer function analysis with an hypothetical event corresponding to the H_y reference model field (H_y measured in the absence of the model structure).

Amplitudes of the field components for some of the traverses over the Vancouver Island model, for a source frequency of 30 kHz, are shown in Fig. 2. As well, the vertical magnetic field component is shown for each of the four field stations. For the model curves, solid lines are used for measurements over the ocean, and dashed lines for measurements over land. As can be seen from the H_z field station values plotted on traverses

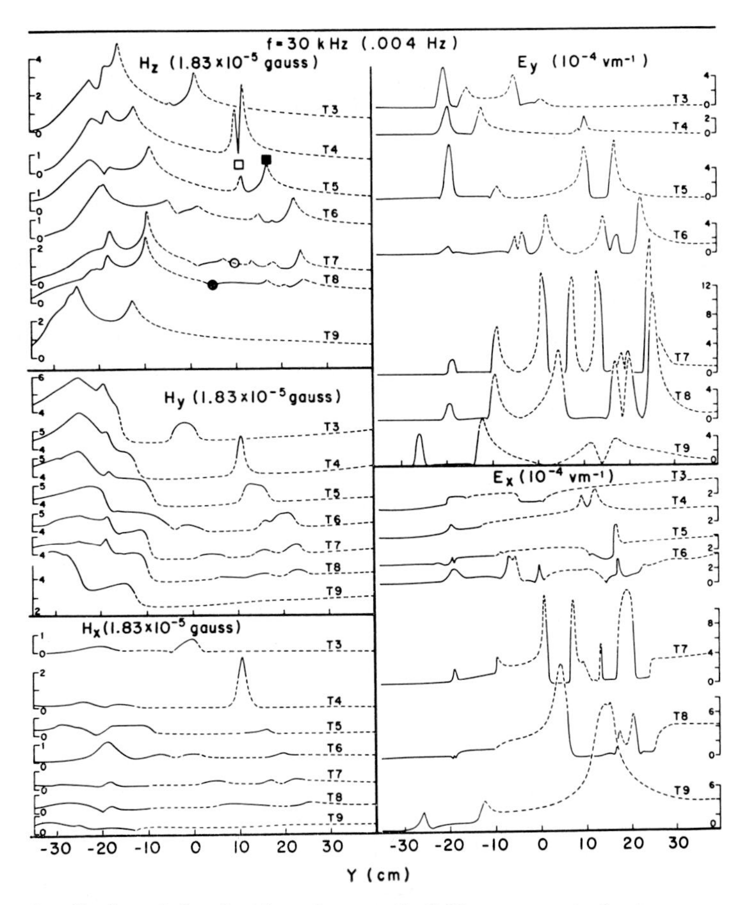

Fig. 2. Amplitudes of the electric and magnetic field components for traverses over the Vancouver Island model for a model frequency of 30 kHz simulating a frequency of 0.004 Hz, and H_z values for four field stations. Full and broken lines correspond to sea and land sections of the traverses respectively (10^{-5} gauss = 1 nT).

T5, T7, and T8, close agreement is found between model and field station H_z results for each of the four stations, with the exception that the Miracle Beach H_z value is greater than the model value. This difference can readily be accounted for in that the Miracle Beach field station is somewhat north of the model traverse T5 where the channel is narrower and the contour of the coast is different. Further, the channel in the model is highly idealized, omitting the various small islands and inlets in the Miracle Beach region of the channel.

For the model frequency of 30 kHz, the seawater channel depth is only of the order of 1/20 of a skin depth and it is unlikely that local induction to the channel could account for the large model H_z anomalies at the island coastline in the channel. To produce the anomalies, conductive current channeling must take place in the ocean channel. Evidence of local channeling as the ocean channel width changes can be seen by considering traverses T4 and T5. The H_z component has two maxima, where T4 intersects the narrow portion of the channel, which are considerably larger than the H_z maximum where T5 crosses the wider channel. The H_y enhancement at the channel crossing is also larger for T4 than for T5. The H_z, H_y, and E_x components all indicate that the current density is greater for the T4 channel crossing than for the T5 crossing. The anomalies in H_x, H_y, and E_y at the T4 channel crossing also show that the geometry of the channel determines the direction of local channeling of telluric current. The channel makes an angle of approximately 45° with the direction of the electric field of the inducing source, leading to an E_y enhancement, and resulting in an almost identical behaviour of both H_x and H_y at the channel crossing.

Evidence of the channeled current spreading out into the continent is seen by considering the behaviour of E_x and E_y for T9 just south of the island. The large E_x anomaly there denotes a large current concentration in the x-direction over the continent. The E_y enhancement on either side of the E_x maximum is indicative of current concentration to the left and to the right. E_x and E_y in this region describe current spreading into the continent. The phases of E_x and E_y (not shown in the present work) also were con-

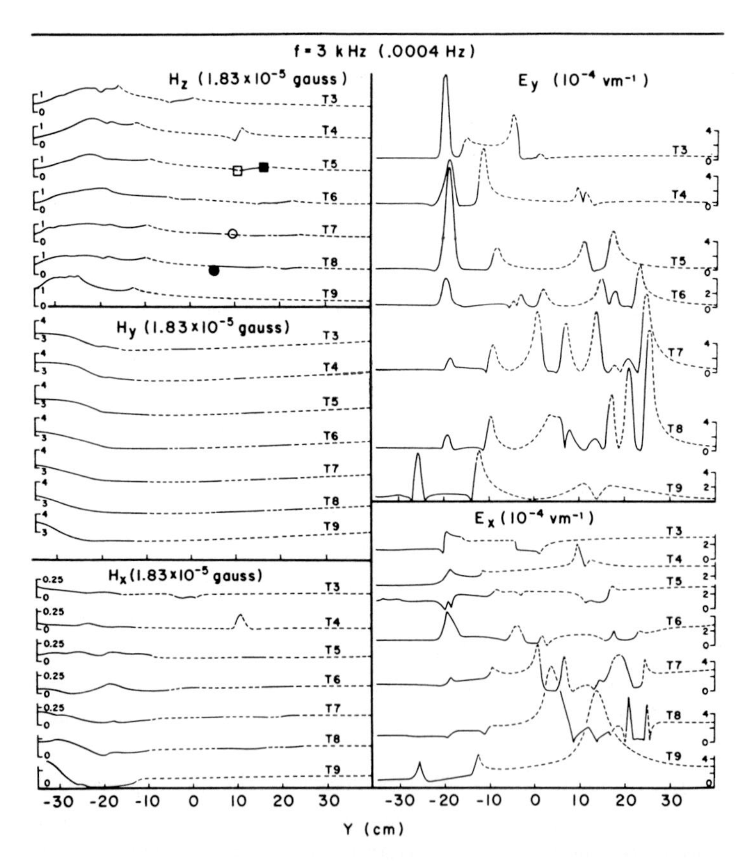

Fig. 3. Amplitudes of the electric and magnetic field components for traverses over the Vancouver Island model for a model frequency of 3 kHz simulating a frequency of 0.0004 Hz, and H_z values for four field stations. Full and broken lines correspond to sea and land traverses respectively (10^{-5} gauss = 1 nT).

sistent with current spreading into the continent.

Amplitudes of the model field components for a model frequency of 3 kHz, simulating a 40–50 minute period, and the H_z component for each of the four field stations is shown in Fig. 3. Comparing the results of Fig. 2 and Fig. 3, it is seen that the magnetic field anomalies are much smaller for 3 kHz than for 30 kHz, while the electric field anomalies are larger for 3 kHz. From the anomalies in H_z and H_x for the T4 channel crossing, it is seen that current channeling present at 30 kHz is present at 3 kHz as well. For 3 kHz, as well as for 30 kHz, E_x and E_y for T9 are consistent with telluric current spreading into the continent at the lower end of the channel. Hence the features of local current channeling and current diffusing into the continent are present at both frequencies.

In general, the field station H_z values show close agreement with the model results, demonstrating that for the periods studied, the electromagnetic response for Vancouver Island can be explained by a simple model.

REFERENCES

Bailey, R. C., R. N. Edwards, G. D. Garland, R., Kurtz, and D. Pitcher, Electrical conductivity studies over a tectonically active area in Eastern Canada, *J. Geomag. Geoelectr.*, **26**, 125–146, 1974.

Dosso, H. W., A plane-wave analogue model for studying electromagnetic variations, *Can. J. Phys.*, **44**, 67–80, 1966a.

Dosso, H. W., Analogue model measurements for electromagnetic variations near a coastline, *Can. J. Earth Sci.*, **3**, 917–936, 1966b.

Dosso, H. W., A review of analogue model studies of the coast effect, *Phys. Earth Planet. Inter.*, **7**, 294–302, 1973.

Cochrane, N. A. and R. D. Hyndman, Magnetotelluric and magnetovariational studies in Atlantic Canada, *Geophys. J. R. Astr. Soc.*, **39**, 385–406, 1974.

Edwards, R. N., L. K. Law, and A. White, Geomagnetic variations in the British Isles and their relationship to electrical currents in the ocean and shallow seas, *Phil. Trans. Roy. Soc.*, **270**, 289, 1971.

Whitham, K. and F. Anderson, Magnetotelluric experiments in northern Ellesmere Island, *Geophys. J.*, **10**, 317–345, 1965.

Induction in Arbitrarily Shaped Oceans II: Edge Correction for the Case of Infinite Conductivity

R. C. HEWSON-BROWNE* and P. C. KENDALL**

**Department of Applied Mathematics and Computing Science, The University, Sheffield, England*
***Department of Mathematics, University of Keele, Keele, Staffordshire, England*

(Received September 10, 1977; Revised January 23, 1978)

Analytic edge corrections are developed suitable for matching with approximate solutions (derived otherwise) for perfectly conducting oceans. By comparison with the solution of a particular problem, it is verified that the matched solutions are good all the way to the edge of the idealized oceans. They reproduce accurately the well known singularity on the coastline, and this property is essential if the solutions are in turn to match those for finite conductivity. Often an appreciable (oscillating) magnetic field passes beneath the "ocean", and this should perhaps be included in future calculations involving ocean edge effects.

1. Introduction

In paper I (HEWSON-BROWNE and KENDALL, 1978) we developed an approximate method of solution for the problem of induction in an infinitely conducting idealized ocean (on a sphere of radius a) above a concentric sphere of radius b, also of infinite conductivity. The configuration is shown in Figs. 1 and 2. The three regions 1, 2, and 3 are chosen as shown, where region 3 lies between the outer and inner spheres. If P denotes

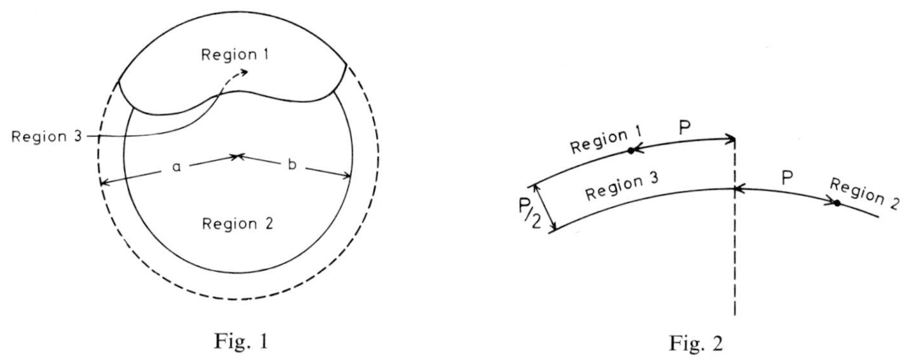

Fig. 1. Showing the idealized ocean with the perfectly conducting sphere simulating the mantle. Region 1 lies just above the ocean, region 2 just above the perfect conductor where it is not covered by the ocean, and region 3 lies between the ocean and the perfect conductor.

Fig. 2. Details near the coastline. The image currents lie approximately at depth P, where $P/2$ is the depth below the earth's surface of the perfect conductor.

the Price depth, $2(a-b)$, the method holds provided that (i) P is small compared with a, (ii) the radius of curvature of the shoreline is large compared with P, (iii) the spatial variation of the source field is large compared with P. The magnetic potentials in these regions are taken to be Ω_1, Ω_2, and Ω_3 derived as follows.

Ω_1: the magnetic potential of the field which arises when the given source field is applied to an infinitely conducting sphere of radius a.

Ω_2: the same, for a sphere of radius b.

Ω_3: the solution of $\nabla_s^2\Omega_3=0$ in region 3 with $\Omega_3=\Omega_2$ at its boundary. The surface Laplacian is denoted by ∇_s^2.

Note that region 3 is assumed thin so that, apart from any time-dependence, Ω_3 will be a function only of latitude θ and longitude ϕ. Thus Ω_3 satisfies

$$\sin\theta\frac{\partial}{\partial\theta}\left(\sin\theta\frac{\partial\Omega_3}{\partial\theta}\right)+\frac{\partial^2\Omega_3}{\partial\phi^2}=0\,. \tag{1}$$

The physical basis of the method is that the magnetic field component parallel to the coastline is the component most likely to be continuous in a first order matching process across the coastline. On the coastline we put $\Omega_3=\Omega_2$ (rather than $\Omega_3=\Omega_1$) on empirical grounds: it gives slightly better results on comparison with the solution of a particular problem. As can be seen, the method is easy to use. In paper I (1978) we verified its accuracy for the spherical cap with an oscillating transverse magnetic field. We observed that, for obvious reasons, the approximation begins to fail at a distance of order P from the coastline.

Figure 2 shows the situation close to the edge of the ocean. The perfectly conducting sphere lies at depth $P/2$ and the images of the electric current in the ocean lie at depth P. The configuration has become basically two-dimensional, as in Fig. 38 of SCHMUCKER (1970). This is the physical basis of the matching technique. We use the nomen-clature "outer" solution to denote the approximation at great distances L from the coastline; ideally $P \ll L$. As an outer solution we use the one obtained by the method we have just outlined. The "inner" solution denotes the approximation (which we intend to derive) at distances L from the coastline, where $L \ll a$. In this latter case, L may exceed P, showing that the inner solution overlaps with the outer solution. This notation is consistent with the terminology of matched asymptotic expansions (VAN DYKE, 1964). The inner problem, shown in Fig. 3, is a two-dimensional problem soluble by complex mapping techniques. We match the horizontal component of the magnetic induction vector at right angles to the coastline at three points, as shown. This requires a solution with three assignable free constants.

2. Solution of the Inner Problem

Choose axes Oxy with origin at the edge (Fig. 3) and let $z=x+iy$. Denote by w the complex potential such that $\Omega=\mathrm{Re}(w)$. Then the components of the induction vector (B_x, B_y) in this plane are given by

$$-B_x+iB_y=\mathrm{d}w/\mathrm{d}z\,. \tag{2}$$

We map the z-plane onto the ζ-plane by the transformation

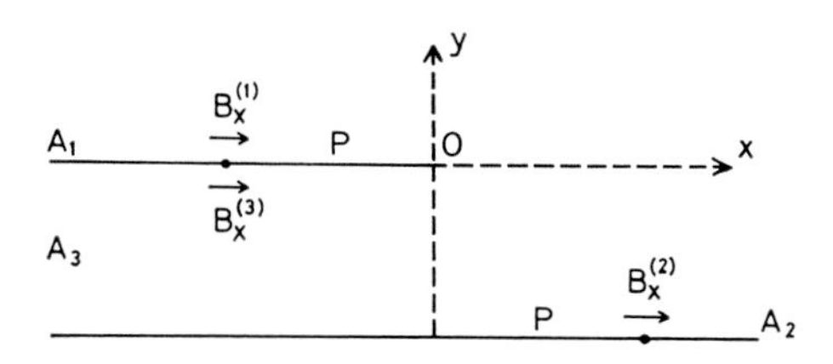

Fig. 3. The flat version of Fig. 2. Points A_1, A_2, and A_3 lie at infinity. The matching points lie at a distance P from the vertical plane through O at right angles to Ox. The perfectly conducting sphere is represented by the line A_2A_3, the ocean by the line A_1OA_3.

$$z=(P/2\pi)\{\zeta+\log(\zeta-1)-i\pi\}\,. \tag{3}$$

This maps the line $A_1OA_3A_2$ representing the ocean and underlying conductor onto the whole of the real ζ-axis. On A_1O, $\zeta=-\lambda_1'$, (say) where λ_1' is real and positive, so $z=d_1\exp(i\pi)$, where d_1 is real and positive and

$$2\pi d_1/P=\lambda_1'-\log(1+\lambda_1')\,. \tag{4}$$

Likewise, on OA_3, $z=d_3\exp(-i\pi)$ and $\zeta=\lambda_3$ $(0\leq\lambda_3\leq 1)$, where

$$2\pi d_3/P=-\{\lambda_3+\log(1-\lambda_3)\} \tag{5}$$

and on A_3A_2, $z=d_2-iP/2$ and $\zeta=\lambda_2$ $(1\leq\lambda_2<\infty)$, where

$$2\pi d_2/P=\lambda_2+\log(\lambda_2-1)\,. \tag{6}$$

Equations (4) to (6) provide the links from the ζ-plane to the physical configuration of Fig. 3. Moreover, near $\zeta=0$, Eq. (3) gives $z\sim-\zeta^2(P/2\pi)$. After some analysis it emerges that dw/dz has a simple pole at $\zeta=0$ in the ζ-plane. Using (2) it follows that the simplest inner solution with three disposable constants is

$$B_x-iB_y=\alpha_1\zeta^{-1}+\alpha_2+\alpha_3\zeta\,, \tag{7}$$

where, for this approximation, we ignore higher order terms. It will also appear that the constants α_1, α_2, and α_3 are real.

3. Matching to the Outer Problem

We equate the values of B_x at $d_1=d_3=d_2=P$ to the values $B_x^{(1)}$, $B_x^{(3)}$, and $B_x^{(2)}$ supplied by the solution of the outer problem. The *rationale* behind this is that in the inner problem the edge effect rapidly disappears at distances exceeding $P/2$ from O; thus twice this distance seems to be a safe compromise. The corresponding values λ_1', λ_2, and λ_3 may be obtained from Eqs. (4), (5), and (6) by iteration. Thus,

$$\{\lambda_1'\}_{n+1}=2\pi d/P+\{\log(1+\lambda_1')\}_n\,, \tag{8}$$

yields $\lambda_1'=8.5385$ at $d_1=P$. Also,

$$\{\lambda_3\}_{n+1}=1-\{e^{-2\pi d_3/P-\lambda_3}\}_n\,, \tag{9}$$

gives $\lambda_3=0.99931$ at $d_3=P$; and

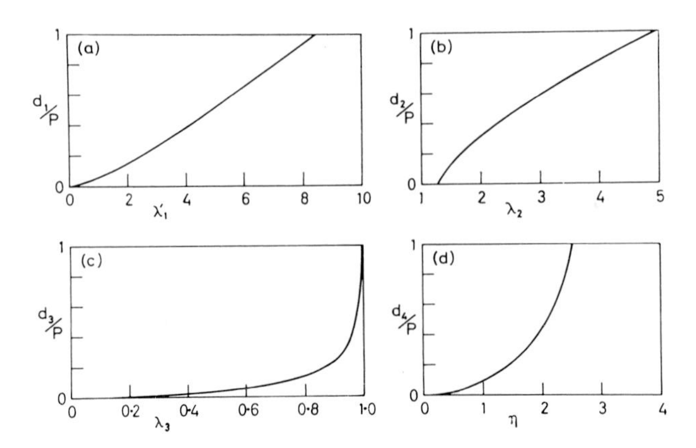

Fig. 4. Graphs of d_1/P, d_2/P, d_3/P, and d_4/P as functions of λ_1', λ_2, λ_3, and λ_4.

$$\{\lambda_2\}_{n+1}=2\pi d_2/P-\{\log(\lambda_2-1)\}_n \tag{10}$$

converges to $\lambda_2=4.9177$ at $d_2=P$. The iterations may be started from approximate values given by Figs. 4(a), (b), and (c). Substituting these values into (7) with $\zeta=\lambda_1$ $(=-\lambda_1')$, λ_3, λ_2 gives three equations for α_1, α_2, and α_3, namely,

$$\left.\begin{aligned}\alpha_1\lambda_1^{-1}+\alpha_2+\alpha_3\lambda_1&=B_x^{(1)}\\ \alpha_1\lambda_2^{-1}+\alpha_2+\alpha_3\lambda_2&=B_x^{(2)}\\ \alpha_1\lambda_3^{-1}+\alpha_2+\alpha_3\lambda_3&=B_x^{(3)}\,.\end{aligned}\right\} \tag{11}$$

Approximately, using the previously derived values,

$$\begin{pmatrix}\alpha_1\\ \alpha_2\\ \alpha_3\end{pmatrix}=\begin{pmatrix}-0.3269 & -0.7958 & 1.1119\\ 0.3937 & 0.7032 & -0.0959\\ -0.0665 & 0.0933 & -0.0265\end{pmatrix}\begin{pmatrix}B_x^{(1)}\\ B_x^{(2)}\\ B_x^{(3)}\end{pmatrix}. \tag{12}$$

Equations (11) were not found to be well-conditioned and the formal solution was used, namely,

$$\left.\begin{aligned}\alpha_1&=\lambda_1\lambda_2\lambda_3\sum_{p=1}^{3}\Delta_p B_x^{(p)}\\ \alpha_3&=\sum_{p=1}^{3}\lambda_p\Delta_p B_x^{(p)}\\ \alpha_2&=-(\lambda_1+\lambda_2+\lambda_3)\alpha_3+\sum_{p=1}^{3}\lambda_p^2\Delta_p B_x^{(p)}\,,\end{aligned}\right\} \tag{13}$$

where $\Delta_1=1/\{(\lambda_3-\lambda_1)(\lambda_2-\lambda_1)\}$, $\Delta_2=1/\{(\lambda_1-\lambda_2)(\lambda_3-\lambda_2)\}$, and $\Delta_3=1/\{(\lambda_2-\lambda_3)(\lambda_1-\lambda_3)\}$.

4. Solution of a Particular Problem (for Comparison)

Consider the spherical cap $0\leq\theta\leq\pi/2$ with a transversely applied magnetic field, B_0 uniform at infinity. Then the source potential is

$$\Omega_\infty=B_0 r\sin\theta\cos\phi\,. \tag{14}$$

We take $b=0.9a$, giving $P/a=0.2$ and $P/b=0.\dot{2}$. The matching points then lie on the circles $\theta=\pi/2-0.2$ ($r=a$) and $\theta=\pi/2+0.\dot{2}$ ($r=b$) which are both close to the equator. Using the procedure outlined in the introduction gives the following approximate solutions. In region 1 (where $0\leq\theta<\pi/2$, just above the cap)

$$\Omega_1=B_0(r+\tfrac{1}{2}a^3/r^2)\sin\theta\cos\phi\,. \tag{15}$$

In region 2 (where $\pi/2<\theta\leqq\pi$, just above the sphere $r=b$)

$$\Omega_2=B_0(r+\tfrac{1}{2}b^3/r^2)\sin\theta\cos\phi\,. \tag{16}$$

In region 3 (where $0\leq\theta<\pi/2$, just below the cap), solving Eq. (1) and using the outer matching condition $\Omega_3=\Omega_2$ on the circle $r=a$, $\theta=\pi/2$, gives

$$\Omega_3=B_0(1+k^3/2)a\tan\theta/2\cos\phi\,. \tag{17}$$

We note that the outer matching condition is equivalent to a demand that the horizontal magnetic induction vector, $-a^{-1}\partial\Omega/\partial\phi$, parallel to the coastline (where $\sin\theta=1$) be continuous between regions 2 and 3. Alternatively, we could ensure that B_ϕ is continuous between regions 1 and 2 at the edge. This is equivalent to making the normal component of electric current on the coastline vanish. The first choice of condition does not give this strictly but is merely consistent with the order of the approximation and gives slightly better results away from the coast. As the solution is separable we are in this case able to take out a factor $\cos\phi$ and match all the way around the coastline.

From Eqs. (15) and (17) we see that a magnetic field of appreciable strength passes underneath the ocean. Taking $r=a$ and θ small shows that at the pole $\theta=0$ the magnetic field just beneath the ocean is approximately half that just above it.

Returning to the links with the inner solution we see that for insertion on the right hand side of (12) we have

$$\left.\begin{aligned}
B_x^{(1)}/B_0\cos\phi&=-\frac{3}{2}\cos\left(\frac{\pi}{2}-0.2\right)=-0.2980\\
B_x^{(2)}/B_0\cos\phi&=-\frac{3}{2}\cos\left(\frac{\pi}{2}+0.\dot{2}\right)=0.3306\\
B_x^{(3)}/B_0\cos\phi&=-\left(1+\frac{k^3}{2}\right)\frac{1}{2}\sec^2\left(\frac{\pi}{4}-0.1\right)=-1.1384\,.
\end{aligned}\right\} \tag{18}$$

Substituting into Eqs. (12) gives

$$\alpha_1=-1.432 \qquad \alpha_2=0.224 \qquad \alpha_3=0.081\,. \tag{19}$$

These constants yield the components (B_x, B_y) through Eqs. (7) and (3). To derive various physical variables for comparison purposes we proceed as follows.

The surface current component J_ϕ may be obtained by identifying B_x with B_θ and B_y with B_r. Thus if μ_0 is the permeability of free space,

$$\mu_0 J_\phi=B_x^+-B_x^-\,. \tag{20}$$

The only difficult quantity to derive is the magnetic induction field *inland* (at distance d_4 along the x-axis). Then Eq. (3) gives with $\zeta=\xi+i\eta$

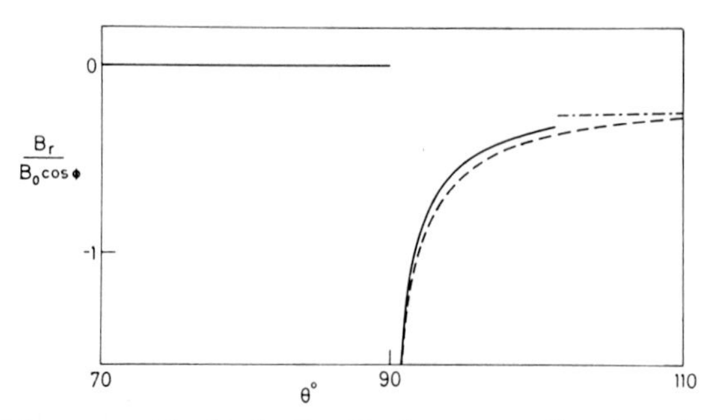

Fig. 5. The radial component of induction B_r shown as a function of colatitude θ near the edge of a spherical cap. For convenience $B_r/(B_0 \cos \phi)$ is shown, where ϕ is the longitude. The (small) discontinuity lies at the matching point.

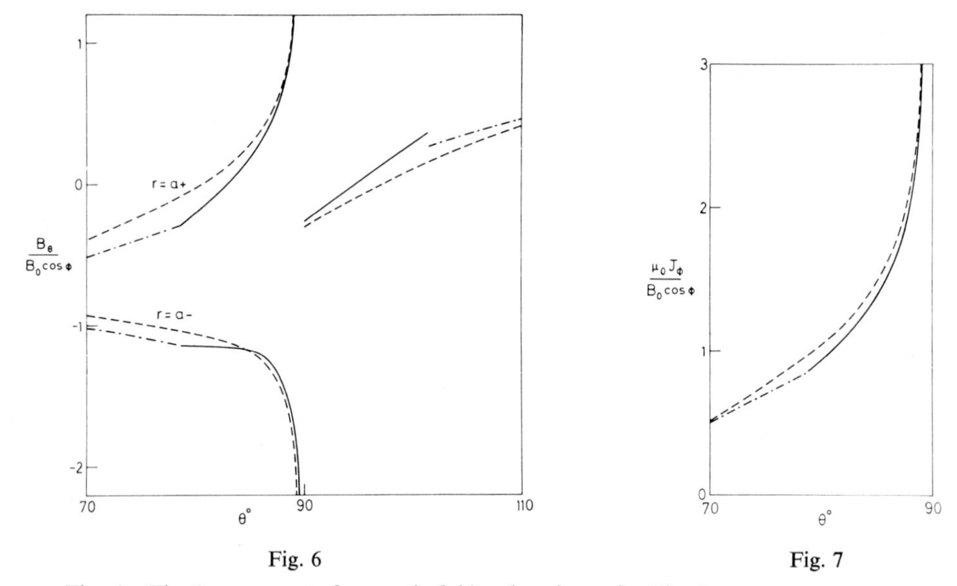

Fig. 6 Fig. 7

Fig. 6. The θ-component of magnetic field; otherwise as for Fig. 5.

Fig. 7. The ϕ-component of electric current just off-shore (the component parallel to the coastline); otherwise as for Fig. 5.

$$2\pi d_4/P = \xi + i\eta + \log(\xi - 1 + i\eta) - i\pi \,. \tag{21}$$

Equating real and imaginary parts gives

$$\xi = 1 - \eta \cot \eta \tag{22}$$

and

$$2\pi d_4/P = 1 - \eta \cot \eta + \log(\eta \operatorname{cosec} \eta) \,, \tag{23}$$

where $\xi \geq 0$ and $0 \leq \eta \leq \pi$. While Eq. (7) gives

$$\left.\begin{aligned} B_x &= \alpha_1 \xi(\xi^2 + \eta^2)^{-1} + \alpha_2 + \alpha_3 \xi \\ \text{and} \quad B_y &= \alpha_1 \eta(\xi^2 + \eta^2)^{-1} - \alpha_3 \eta \,. \end{aligned}\right\} \tag{24}$$

We suggest that Eq. (24) for η in terms of d_4 may be solved iteratively from the relation

$$\{\eta\}_{n+1}=\left\{\tan^{-1}\frac{\eta}{1-(2\pi d_4/P)+\log(\eta/\sin\eta)}\right\}_n . \tag{25}$$

The iterations may be started by using rough values obtained from Fig. 4 (d). As a check the value $\eta=2.5472$ may be obtained for $d_4=P$.

Figures 5, 6, and 7 show the approximate spatial variations of B_r, B_θ, and J_ϕ compared with the Legendre series solution of paper I. We see that, considering the ease of application of the matching technique, agreement is good.

5. Conclusion

The problem of a perfectly conducting ocean of arbitrary shape above an underlying sphere, also of perfect electrical conductivity is soluble by matching techniques, provided that $a-b$ is small compared with a, and also with the radius of curvature of the coastline. Comparison with a Legendre series solution for the case of a hemispherical cap shows good agreement. The matching technique accurately reproduces the well known singularity in J_ϕ on the coastline.

This latter result is important in order to extend these results to the case of finite conductivity. Hewson-Browne (1973) has shown how the finitely conducting case is related to the infinitely conducting case when

$$\gamma\equiv\omega\kappa_0\mu_0(a-b)\gg 1 ,$$

where ω is the frequency, κ_0 a representative surface conductivity, and μ_0 the permeability of free space. As usual in asymptotic expansions, the results obtained from matching are better than expected, even for $\gamma\sim 1$. As this value of γ is achieved at periods of about 12 hr, we see that we may apply the results directly to a wide range of problems. Dr. D. A. Quinney has already produced world curves for an ocean of realistic shape. We hope to report the full calculations in due course, along with the extensions to include finite conductivity.

The method introduced in this paper also provides approximated *analytic* solutions to a considerable number of idealized problems, all with a closely situated underlying perfect conductor. These include oceans of analytic shapes, island problems, problems involving cylinders and plane problems (*inter alia*). In many cases a quick estimate is obtained of the magnetic flux passing beneath the ocean. We have here shown how to do this for a spherical cap.

Finally we note that in providing the connection between the infinitely conducting and finitely conducting problems Hewson-Browne and Kendall (1976) have derived a simplified version of the integral equation for the strip problem derived by Roden (1964). It is through this equation that the link with the finitely conducting problem may be obtained computationally. The magnetic merging calculations of Hewson-Browne and Kendall (1973) give a physical understanding of the near-singular role played by the ocean's edge when the conductivity is finite.

The major part of this paper was included in a paper "First order solution of oceanic induction problems" kindly presented for us at Symposium S14, IAGA, Seattle 1977 by Dr. D. M. Barraclough.

REFERENCES

HEWSON-BROWNE, R. C., Estimates for induced electric currents flowing near the coast, and their associated magnetic fields just inland, *Geophys. J. R. Astr. Soc.*, **34**, 393–402, 1973.

HEWSON-BROWNE, R. C. and P. C. KENDALL, Induction in the oceans explained and analysed in terms of magnetic merging, *Geophys. J. R. Astr. Soc.*, **34**, 381–391, 1973.

HEWSON-BROWNE, R. C. and P. C. KENDALL, Magneto-telluric modelling and inversion in three dimensions, *Acta Geod. Geophys. Mont. Hung.*, **11**, 427–446, 1976.

HEWSON-BROWNE, R. C. and P. C. KENDALL, Some new ideas on induction in infinitely conducting oceans of arbitrary shapes, *Geophys. J. R. Astr. Soc.*, **53**, 431–444, 1978.

RODEN, R. B., The effect of an ocean on magnetic diurnal variations, *Geophys. J. R. Astr. Soc.*, **8**, 375–388, 1964.

SCHMUCKER, U., Anomalies of geomagnetic variations in the Southwestern United States, *Bull. Scripps. Inst. Oceanogr.*, **13**, 1–165, 1970.

VAN DYKE, M., *Perturbation Methods in Fluid Mechanics*, Academic, 1964.

Note added in proof

Since this paper was submitted, BEAMISH *et al.* (1980a, b) have produced world curves for realistic oceans of finite conductivity based on theory presented in paper III of this series (HEWSON-BROWNE, 1978). The results provide an outer solution and are valid away from the coastline. Possible edge corrections for this case have recently been proposed by QUINNEY (1979).

BEAMISH, D., R. C. HEWSON-BROWNE, P. C. KENDALL, S. R. C. MALIN, and D. A. QUINNEY, Induction in arbitrarily shaped oceans IV: *Sq* for a simple case, *Geophys. J. R. Astr. Soc.*, **60**, 435–443, 1980a.

BEAMISH, D., R. C. HEWSON-BROWNE, P. C. KENDALL, S. R. C. MALIN, and D. A. QUINNEY, Induction in arbitrarily shaped oceans V: The circulation of *Sq*-induced currents around land masses, *Geophys. J. R. Astr. Soc.*, 1980b (in press).

HEWSON-BROWNE, R. C., Induction in arbitrarily shaped oceans III: Oceans of finite conductivity, *Geophys. J. R. Astr. Soc.*, **55**, 645–654, 1978.

QUINNEY, D. A., A note on computing coastal edge corrections for induced oceanic electric fields, *Geophys. J. R. Astr. Soc.*, **49**, 119–126, 1979.

The Effect of a Simple Model of the Pacific Ocean on *Sq* Variations

B. A. Hobbs and G. J. K. Dawes

Department of Geophysics, University of Edinburgh, U.K.

(Received December 5, 1977; Revised February 10, 1978)

The problem solved is that of electromagnetic induction in a thin non-uniformly conducting hemispherical shell underlain by, and insulated from, a perfectly conducting sphere Since the shell is assumed thin, only the component of a varying magnetic field normal to the shell induces electric current in the shell. This mode of induction is termed vertical component induction. The conductance and configuration of the surface shell are simple approximations to those of the Pacific Ocean. Calculations are made of the effect of induction by the *Sq* variation field and results are presented for typical "land" and "ocean" regions, and for a profile of stations near the ocean edge. The results from this profile of stations are compared to observations made by Schmucker (1970) near the coast of California. The calculated and observed daily variations of the vertical component of the magnetic field are in qualitative agreement.

1. Introduction

Many measurements of magnetic field variations have been made near coasts bordering large oceans. Measurements near the Pacific coast around California have been described by Schmucker (1970), and Fig. 1 shows a particular example of the data he obtained. Each curve represents the vertical component of the measured magnetic field throughout one day, commencing at local midnight. The stations are arranged from top to bottom in increasing distance inland from the coast. The stations form a profile approximately perpendicular to the coastline. These data show one typical feature of the "coast effect", that is the doubling, approximately, of the vertical magnetic field variations in the short distance of about 300 km. Theoretical aspects of this edge effect, which is a result of the conductivity contrast between land and sea, have been studied for some time by considering especially simple inducing fields and using perfectly or uniformly conducting shells, or shells whose conductivity distribution is specified analytically (see Ashour (1973) for a review). A new method of solution was proposed by Hobbs and Bringnall (1976) which enables solutions to be determined to general thin-shell problems. This present paper concerns an application of the general method to determine to what extent electric currents, induced in the Pacific Ocean by vertical component induction, are responsible for the "coast effect" observed by Schmucker (1970).

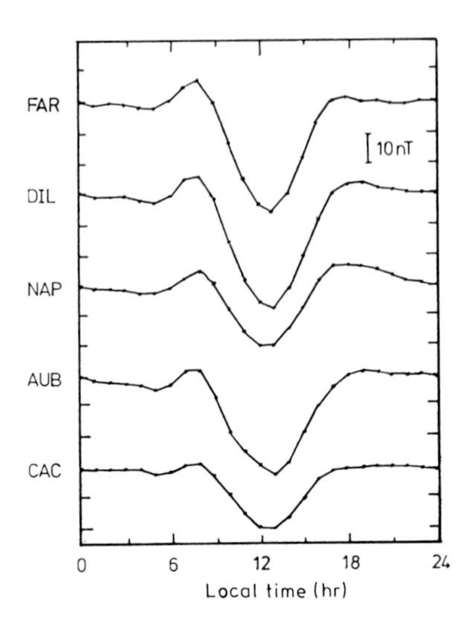

Fig. 1. Enhancement of the diurnal vertical field amplitude near the coast of California. The profile runs from the inland station of Carson City (CAC), through Auburn (AUB), Napa (NAP) and Dillons Beach (DIL) to the Farallon Island (FAR) just off the coast. (Redrawn from SCHMUCKER, 1970).

2. The Model

A thin, non-uniformly conducting hemispherical shell is taken to represent the ocean. A non-conducting hemispherical shell represents the land and this combined surface shell surrounds, and is electrically insulated from, an inner, perfectly conducting sphere which models the conducting part of the mantle. The conductance of the hemispherical shell may vary from the centre of the shell to the edge. In particular, the conductance may decrease to zero rapidly, or slowly, as the "coastline" is approached. The first requirement of the model, therefore, is that this variable conductance should resemble that of the ocean near the observation profile.

A magnetic tape giving values of the conductance of the world's oceans, averaged over 10°, 5°, 2°, and 1° tesserae, has been produced by E. C. Bullard. Figure 2 shows the results from the 1° values on this tape for a large area surrounding California. The conductances are contoured, the zero contour representing the coastline. The other contours stem from smoothed values of the conductance, even so it can be seen that this oceanic region is somewhat complicated. However, it is apparent that the conductance increases rapidly from the coastline and approaches its normal oceanic value in only 4 or 5 degrees longitude. More strikingly, a bathymetric map of the area shows that the 4 km ocean depth contour is reached in only 2 degrees longitude from the coast. What is required is a one-dimensional variation of conductance that approximates this complicated two-dimensional situation. To do this, two different models are derived, one from Fig. 2 and one directly from the bathymetry. Model 1 is obtained by averaging the smoothed conductances in Fig. 2 along lines parallel to the coast. This gives the smoothest profile consistent with Fig. 2. Model 2 is obtained by multiplying the conductivity of sea-water (4 Sm^{-1}) by the ocean depth along a profile perpendicular to the coast near the observation sites. This gives rise to a very steep gradient in conductance. One might expect that Models 1 and 2 would give the minimum and maximum possible coastal effects, respectively. The variation in conductance for these two models is shown in Fig. 3.

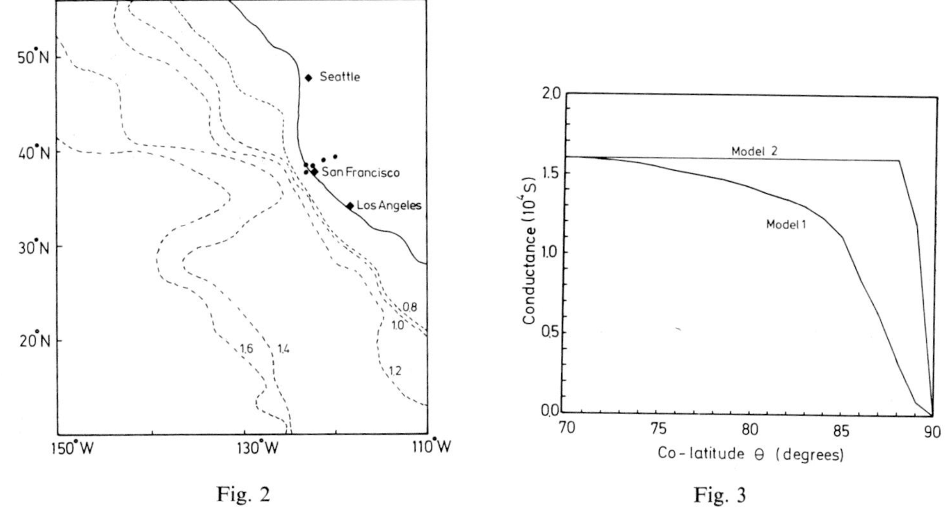

Fig. 2 Fig. 3

Fig. 2. Contours of ocean conductance in units of 10^4 S. The five full circles are the sites of the stations at which the measurements of Fig. 1 were made, running from the inland station CAC to FAR just off the coast.

Fig. 3. The one-dimensional conductance profiles of models 1 and 2. The variation in conductance is shown for $70° < \Theta < 90°$. For $\Theta < 70°$, the conductance has the value 1.6×10^4 S.

3. The Inducing Field

In order to simulate the data shown in Fig. 1, the inducing field is considered to be that of *Sq*, the magnetic potential of which may be written as the spherical harmonic expansion

$$\Omega_{Sq} = a \sum_n \sum_m \sum_k \left(\frac{r}{a}\right)^n \{(AA^m_{nk} \cos m\lambda + AB^m_{nk} \sin m\lambda) \cos k\alpha t$$
$$+ (BA^m_{nk} \cos m\lambda + BB^m_{nk} \sin m\lambda) \sin k\alpha t\} P^m_n(\cos\theta) \tag{1}$$

in a spherical polar coordinate system (r, θ, λ), with suitable coefficients AA, AB, BA, and BB. t is universal time in seconds, $\alpha = 2\pi/86{,}400$ and P^m_n denotes the Schmidt partially normalised associated Legendre function of degree n, order m. It is generally accepted that the important harmonics in this expansion are given by $m=1, 2, 3$; $n=m, m+1$; $k=m$; accordingly, expansion (1) is restricted to these terms. The coefficients used here are those of MALIN (1973) and are listed in Table 1.

Sq analyses yield coefficients for the above expansion in terms of spherical harmonics relative to geographic coordinates—that is the polar axis of the spherical polar coordinate system (r, θ, λ) passes through the geographic pole. The hemispherical shell used here represents the Pacific Ocean reasonably well if the geographic axis passes through its rim. Therefore, as in HOBBS (1971), the problem is made more tractable by rotating the coordinates (r, θ, λ) by $\pi/2$ radians to a new coordinate system (r, Θ, Λ) in which the polar axis passes through the centre of the hemispherical ocean. In the new coordinate system, the

Table 1. Coefficients of the principal spherical harmonic terms in the description of *Sq* (taken from MALIN, 1973).

n, m	AA^m_{nm}	AB^m_{nm}	BA^m_{nm}	BB^m_{nm}
2, 1	11.0	−1.6	−1.5	−13.2
3, 2	−6.3	0.3	0.0	5.6
4, 3	2.1	−0.7	−0.4	−2.1
1, 1	1.5	−1.3	−1.0	0.0
2, 2	0.2	1.5	1.0	0.9
3, 3	0.0	−0.3	−0.7	0.0

Table 2a. Coefficients for the rotation of the surface harmonics (Eq. (2)) associated with $\cos m\lambda$.

n, m	j				
	0	1	2	3	4
1, 1	−1	0	0	0	0
2, 1	0	−1	0	0	0
2, 2	$\sqrt{3}/2$	0	1/2	0	0
3, 2	0	$\sqrt{10}/4$	0	$\sqrt{6}/4$	0
3, 3	$-\sqrt{10}/4$	0	$-\sqrt{6}/4$	0	0
4, 3	0	$-\sqrt{7}/4$	0	−3/4	0
4, 4	$\sqrt{35}/8$	0	$\sqrt{7}/4$	0	1/8

Table 2b. Coefficients for the rotation of the surface harmonics (Eq. (2)) associated with $\sin m\lambda$.

n, m	j				
	0	1	2	3	4
1, 1	0	1	0	0	0
2, 1	0	0	1	0	0
2, 2	0	−1	0	0	0
3, 2	0	0	−1	0	0
3, 3	0	$\sqrt{15}/4$	0	1/4	0
4, 3	0	0	$\sqrt{14}/2$	0	$\sqrt{2}/4$
4, 4	0	$-\sqrt{14}/4$	0	$-\sqrt{2}/4$	0

conductance $\tau=\tau(\Theta)$, the conducting shell occupying $0<\Theta<\pi/2$. On rotation through $\pi/2$, the surface harmonics

$$\left\{\begin{array}{l}\cos m\lambda\\ \sin m\lambda\end{array}\right\}P_n^m(\cos\theta)=\sum_{j=0}^{n}\left\{\begin{array}{l}a^m_{n,j}\cos jA\\ b^m_{n,j}\sin jA\end{array}\right\}P_n^j(\cos\Theta)\,. \tag{2}$$

The coefficients $a^m_{n,j}$ and $b^m_{n,j}$ required for the harmonics necessary to describe the *Sq* field are given in Tables 2a and 2b.

For each of the harmonics listed in Tables 2a and 2b, solution is required to the problem of induction in the non-uniformly conducting shell and the inner conductor. A scrutiny of these tables shows that solution is required to the 14 harmonics whose degrees are $n=1, 2, 3, 4$ and whose orders are $m=0, 1, .., n$. When these solutions have all been determined, they may be synthesized to perform the rotation (2), and further synthesized to form the sum (1), with the *Sq* coefficients *AA*, *AB*, *BA*, *BB* from Table 1.

4. Details of the Calculations

The method of calculation is described fully in HOBBS and BRINGNALL (1976). Analytic continuation is used to enable the "high frequency" parts of the expansion (1) to be dealt with. There are two improvements to that work. First, the conductance of the hemispherical shell is now variable, and can be specified at 5° or 1° intervals. All the calculations in this paper were made with 69 grid points along a "meridian" from $\Theta=0°$ to $\Theta=180°$. The grid points were at intervals of 5° from $\Theta=0°$ to $\Theta=70°$, and also from $\Theta=110°$ to $\Theta=180°$. For $\Theta=70°$ to $\Theta=110°$, grid points were specified at intervals of 1°. This refined grid extends over and beyond the range of interest centred on the ocean/continental boundary. The second improvement is one of accuracy. When the solution had been obtained, it was found that its accuracy could be improved by applying a "low frequency" method (HOBBS, 1971) with this solution as starting value. Two or three iterative cycles reduced the accuracy, as defined in HOBBS and BRINGNALL (1976, p. 538), to less than 0.07% for all harmonics and periods used. This last step is not a necessary one, but is rather esoteric. The values of the current streamline functions and associated magnetic fields were hardly altered during this iterative "improvement", but the accuracy, as formulated in HOBBS and BRIGNALL (1976), is susceptible to this small change.

5. Results

It is convenient to discuss the solution to the problem of induction in this system of conductors by considering three distinct regions. Region A is the continental region well removed from the coast, region B is the oceanic region well removed from the coast, and region C is the immediate neighbourhood of the ocean/continent boundary.

Region A is above a perfect conductor. Accordingly no phase change would be expected between the inducing and induced fields. The largest component of the *Sq* inducing field is that of the P_2^1 harmonic, the phase of which is about 0.5 hr. Thus over region A, the *Z*-variation should reach its maximum amplitude about 0.5 hr before local noon. A typical result for region A (using model 1 or model 2) agrees with this calculation and is shown in Fig. 4a, in the form of the total *Z*-variation throughout one day commencing at local midnight.

The result for a point in region B should be similar to the result for a uniformly conducting surface shell occupying the whole surface $r=a$. From HOBBS (1971), one should expect a phase change due to such an ocean of

$$-\arctan\left\{\frac{\mu a\omega(1-.9^{2n+1})}{(2n+1)\rho}\right\}$$

where μ, ω, and ρ are permeability, frequency and reciprocal conductance, respectively. For the P_2^1 harmonic, and for the value of ρ corresponding to the Pacific Ocean, the phase change is approximately -2.5 hours. Hence over the sea in region B, the maximum *Z*-variation should occur at about $(12.00-0.5+2.5)$ hr local time, i.e. around 14 hr. The

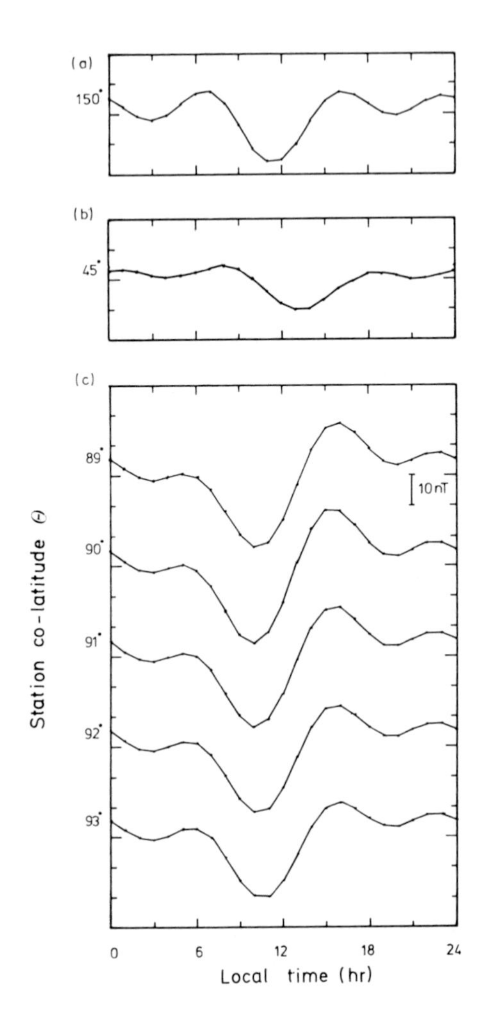

Fig. 4. The calculated amplitude of the total vertical magnetic field due to induction by *Sq*. (a) An inland station. (b) An oceanic station. (c) A profile of five stations perpendicular to the coast. $\Theta=93°$ is inland, and corresponds to the station CAC, $\Theta=89°$ is just off the coast and corresponds to the station FAR.

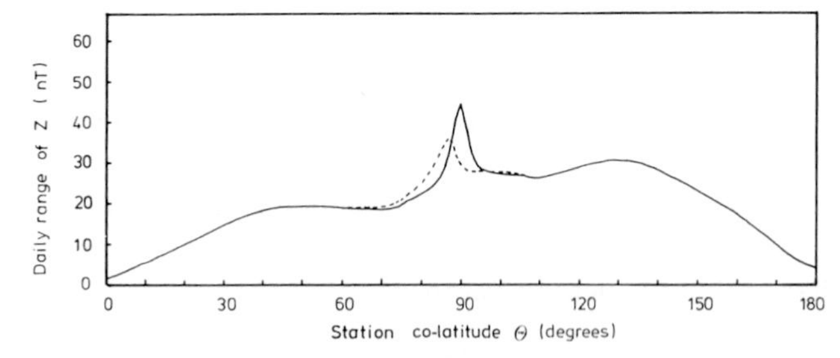

Fig. 5. The daily range of the vertical component of the total magnetic field along the $\Lambda=52°$ meridian, for $0<\Theta<180°$. Dotted line—calculations using model 1; solid line—calculations using model 2.

results are in accord with this value and a typical daily variation is shown in Fig. 4b. Again this result is similar for either model 1 or model 2.

In region C, a profile of stations is considered, the profile being perpendicular to the coast at co-latitude $\Theta=52°$. In the (r, Θ, Λ) coordinate system, for $\Lambda=52°$ the stations are located at $\Theta=89°$, 90°, 91°, 92°, and 93°, these correspond approximately to the observation sites in Fig. 2. The seaward station, Farallon Islands (FAR), corresponds to $\Theta=$ 89°, the most inland station, Carson City (CAC), is represented by $\Theta=93°$. The calculated daily variation at each of these stations now depends on the variation of conductance near the ocean edge. For the smoother conductance, model 1, the *Z*-variation is fairly similar at each of the five stations. For model 2, there is a marked difference in the *Z*-variations, the daily range increasing considerably from the inland to the coastal station. For these two models, Fig. 5 shows the daily range for all "stations" from the centre of the ocean, $\Theta=0$, to the centre of the continent, $\Theta=180°$. The coastal effect can be seen clearly in Fig. 5 as the sharp peak in the otherwise smooth daily range curve. The *Z*-variations are only enhanced over an angular distance of about 4°. For model 1, the enhancement is rather small, and is centred at $\Theta=86°$ (where the conductance has reached about half its normal oceanic value). Thus the stations $89°<\Theta<93°$ show little coastal effect. The enhancement is more pronounced for model 2, and is centred at the coast, $\Theta=90°$. The daily range there is about 45 nT, compared to about 30 nT inland. It is therefore possible that calculations using model 2 could more closely resemble the observations, and Fig. 4c shows the daily variation of the total Z component for the 5 stations of the chosen profile ($\Theta=89°, \ldots, 93°$). The calculations shown in Fig. 4c are to be compared directly with the observations shown in Fig. 1.

Such a comparison shows qualitative agreement. The variations are not too dissimilar in form, and the range increases considerably from the inland station to the coast. The calculations show an earlier phase, and an asymmetry in the positive values near 6 hr and 15 hr local time. It should be noted that the observations were made over a period of about 6 months in 1961, whereas the *Sq* inducing field used in the calculations was derived from 100 observatories operating during 1957/1960—the inducing fields may therefore be somewhat different. If one adds to this the simplicity of the model, a hemispherical shell underlain by a perfect conductor, it is perhaps unreasonable to expect any better argreement than that obtained. The calculation would then point to induction by the vertical component of the inducing field to be the mechanism responsible for the observed coastal effect.

There are two reservations to be made. The first is that to obtain the required increase in the daily range, the more severe conductance profile has had to be used. I think however that this is a reasonable situation. According to bathymetric maps of the area, the deep ocean is reached in only 1° or 2° longitude from the coast in the region of the profile. It is perhaps inappropriate to "smooth" the conductances, therefore. The second reservation is with regard to the phase change along the profile. The observations, Fig. 1, show no phase change, with a peak amplitude around local noon. The calculations, Fig. 4c, exhibit a phase change of around 1 hr inland, to about 2 hr at the coast. Most of this is due to the phase of the inducing field—there remains a small shift in the peak value across the profile in the calculations that does not appear to be present in the observations. It is hoped to shed further light on this problem when calculations are made using a surface shell more representative of the ocean/continent distribution.

6. Conclusions

Detailed calculations have been made of the effect of vertical component induction in a simple thin-shell model of the Pacific Ocean, in the presence of an underlying perfectly conducting sphere representing the conducting part of the mantle. Results are presented for "land", "ocean", and "coastal" regions. Over land and ocean regions well removed from the coast, the magnetic field variations are the same as if the land or ocean, respectively, occupied the complete surface. The most notable effect of the ocean is the well-known edge, or coastal anomaly, in which the vertical field variations are enhanced near the coast. The calculations show that for this mode of induction the enhancement occurs over an angular distance of about 4° (corresponding to 440 km). The calculated daily variations at 5 stations near the ocean edge are in qualitative agreement with observations made by SCHMUCKER (1970), showing that the vertical component induction mode does explain part of the observed anomalous behaviour. The differences between the calculated and observed magnetic fields are presumably to be explained by the horizontal component induction mode, and for that a "thick" shell, permitting vertical current flow, would have to be considered. On the basis of results presented here, it seems worthwhile to seek solutions to the vertical component induction problem using a more representative thin surface shell. Work is therefore in progress on this problem.

This work was supported by the Natural Environment Research Council under grant GR3/2705 by means of which one of us (GJKD) was funded.

REFERENCES

ASHOUR, A. A., Theoretical models for electromagnetic induction in the oceans, *Phys. Earth Planet. Inter.*, **7**, 303–312, 1973.

HOBBS, B. A., The calculation of geophysical induction effects using surface integrals, *Geophys. J. R. Astr. Soc.*, **25**, 481–509, 1971.

HOBBS, B. A. and A. M. M. BRIGNALL, A method for solving general problems of electromagnetic induction in the oceans, *Geophys. J. R. Astr. Soc.*, **45**, 527–542, 1976.

MALIN, S. R. C., Worldwide distribution of geomagnetic tides, *Phil. Trans. Roy. Soc. Lond.*, **274**, 551–594, 1973.

SCHMUCKER, U., Anomalies of geomagnetic variations in the Southwestern United States, *Bull. Scripps Inst. Oceanography, Univ. Calif.*, Vol. 13, 1970.

Electromagnetic Induction at a Model Ocean Coast[†]

G. FISCHER,* P.-A. SCHNEGG,* and K. D. USADEL**

**Observatoire Cantonal, Neuchâtel, Switzerland*
***Gesamthochschule, Fachbereich, Duisburg, GFR*

(Received November 28, 1977; Revised February 2, 1978)

The electromagnetic response of an ocean coast model to a vertically incident monochromatic plane wave is studied. The model consists of a perfectly conducting half-plane (the model ocean) resting at the surface of a good conductor (the model earth). Values of electric and magnetic field components at the surfaces of land and ocean, as well as at the ocean floor, are given in tabular form for both *E*- and *H*-polarization induction.

We have calculated the *E*-polarization electromagnetic response to a vertically incident monochromatic plane wave of a two-dimensional ocean coast model consisting of a perfectly conducting half-plane (the model ocean) resting at the surface of a conductor (the model earth) with which it is in good contact. This model, as well as the coordinate system chosen, is shown in Fig. 1. Our method of solution is analytical and strictly rigorous up to the point where the electric surface field is expressed in terms of an integral equation. This equation is then solved numerically. A detailed description of this work has been published elsewhere (FISCHER *et al.*, 1978).

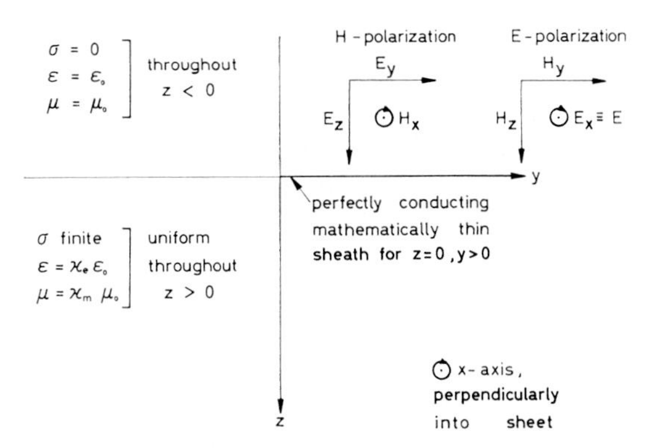

Fig. 1. Cross-section of the conductive two-dimensional structure chosen as model of an ocean coast. The mathematically thin perfect conductor at ($z=0$, $y>0$) represents the ocean. A monochromatic plane wave is incident vertically on this structure. For *H*-polarization induction the only field components concerned are E_y, E_z, and H_x. With *E*-polarization the components involved are E_x, H_y, and H_z.

[†] The authors wish to dedicate this article to Professor Jean Rossel on the occasion of his 60th birthday.

Since there is at present a great deal of interest in comparing the predictions of different numerical methods, as well as those of analog laboratory scale models, with the results of analytical calculations, we present our results in the form of a table of numerical data. To facilitate this kind of comparison we also present a table of similar data for H-polarization induction, derived from an entirely analytical solutions recently published by BAILEY (1977) and NICOLL and WEAVER (1977). These Tables 1 and 2 refer to field component values at the $z=0$ surface of the model of Fig. 1, in the quasi-static ($\omega\,\varepsilon/\sigma \ll 1$) and non-magnetic ($\mu \equiv \mu_0$) limit. Distances along the surface (y-coordinate) are expressed in skin-depths δ,

Table 1. Values of electric and magnetic fields at the surfaces of land and ocean ($z=-0$), as well as at the ocean floor ($y>0$, $z=+0$), for E-polarization induction in the model structure of Fig. 1.

y	ReE_x $z=\pm0$	ImE_x $z=\pm0$	ReH_y $z=-0$	ImH_y $z=-0$	ReH_y $z=+0$	ImH_y $z=+0$	ReH_z $z=\pm0$	ImH_z $z=\pm0$
$-\infty$	1	1	1	0	1	0	0	0
-8	.9899	1.0036	.981	.023	.981	.023	.001	.003
-7	.9828	1.0053	.972	.032	.972	.032	.001	.004
-6	.9735	1.0071	.959	.044	.959	.004	.001	.006
-5	.9601	1.0100	.941	.062	.941	.062	.002	.008
-4	.9382	1.0151	.911	.090	.911	.900	.003	.015
-3	.8961	1.0229	.851	.136	.851	.136	.004	.030
-2	.8031	1.0217	.709	.201	.709	.201	$-.009$	.068
-1	.5925	.9339	.360	.218	.360	.218	$-.104$	.153
$-.8$	.5264	.8836	.244	.196	.244	.196	$-.151$	.179
$-.6$	.4489	.8102	.102	.150	.102	.150	$-.222$	.210
$-.4$	.3568	.7002	$-.071$	.070	$-.071$	.070	$-.340$	.253
$-.2$	.2414	.5224	$-.302$	$-.075$	$-.302$	$-.075$	$-.592$	.337
$-.1$	.1643	.3759	$-.436$	$-.204$	$-.436$	$-.204$	$-.942$	.459
-0	0	0	-0.57	0.00	-0.57	0.00	$-\infty$	$+\infty$
$+0$	0	0	$+\infty$	$-\infty$	$-\infty$	$+\infty$	0	0
$+.1$	0	0	1.520	$-.305$	$-.253$	.273	0	0
.2	0	0	1.374	$-.231$	$-.128$	.188	0	0
.4	0	0	1.245	$-.163$	$-.037$	.105	0	0
.6	0	0	1.184	$-.128$	$-.0057$	.063	0	0
.8	0	0	1.148	$-.107$	$+.0057$	.039	0	0
1	0	0	1.124	$-.092$	.0093	.024	0	0
2	0	0	1.069	$-.056$	.0043	.0005	0	0
3	0	0	1.048	$-.041$	.0006	$-.0008$	0	0
4	0	0	1.037	$-.032$	$-.0001$	$-.0002$	0	0
5	0	0	1.030	$-.027$	$-.00005$	-0	0	0
6	0	0	1.025	$-.023$	-0	$+0$	0	0
7	0	0	1.022	$-.020$	$+0$	$+0$	0	0
8	0	0	1.019	$-.018$	$+0$	$+0$	0	0
$+\infty$	0	0	1	0	0	0	0	0

The electric field parameters are normalized to the value of ReE_x at $y=-\infty$. All magnetic field components are normalized to ReH_y at $y=-\infty$. Coordinate y is expressed in skin depths. These data are for the quasi-static and non-magnetic limit, and were derived from FISCHER *et al.* (1978).

$$\delta = \left(\frac{2}{\omega \mu_0 \sigma} \right)^{1/2}. \tag{1}$$

Real (in phase) and imaginary (in quadrature) parts of electric fields are normalized to the

Table 2. Values of electric and magnetic fields at the surfaces of land and ocean ($z=-0$), at well as at the ocean floor ($y>0$, $z=+0$) for H-polarization induction in the model structure of Fig. 1.

y	ReE_y $z=\pm 0$	ImE_y $z=\pm 0$	ReE_z $z=+0$	ImE_z $z=+0$	ReH_x $z=+0$	ImH_x $z=+0$
$-\infty$	1	1	0	0	1	0
-4	.9998	1.0004	0	0	1	0
-3	.9982	1.0002	0	0	1	0
-2	.9952	.9932	0	0	1	0
-1.5	.9985	.9807	0	0	1	0
-1	1.0234	.9544	0	0	1	0
$-.8$	1.0518	.9389	0	0	1	0
$-.6$	1.1073	.9220	0	0	1	0
$-.5$	1.1545	.9146	0	0	1	0
$-.4$	1.2261	.9108	0	0	1	0
$-.3$	1.3420	.9162	0	0	1	0
$-.2$	1.5551	.9481	0	0	1	0
$-.1$	2.0775	1.0827	0	0	1	0
$-.08$	2.2961	1.1511	0	0	1	0
$-.06$	2.6207	1.2599	0	0	1	0
$-.04$	3.1726	1.4575	0	0	1	0
$-.02$	4.4347	1.9389	0	0	1	0
-0	$+\infty$	$+\infty$	0	0	1	0
$+0$	0	0	$+\infty$	$+\infty$	1	0
.02	0	0	4.3311	1.6933	.8254	$-.0710$
.04	0	0	3.0247	1.1134	.7540	$-.0981$
.06	0	0	2.4381	.8425	.7000	$-.1174$
.08	0	0	2.0836	.6736	.6550	$-.1325$
.1	0	0	1.8381	.5539	.6159	$-.1447$
.2	0	0	1.2056	.2352	.4688	$-.1817$
.3	0	0	.9036	.0840	.3647	$-.1969$
.4	0	0	.7111	$-.0052$	.2845	$-.2005$
.5	0	0	.5722	$-.0616$	.2207	$-.1970$
.6	0	0	.4652	$-.0978$	.1690	$-.1888$
.8	0	0	.3095	$-.1335$	.0926	$-.1651$
1	0	0	.2027	$-.1409$	.0420	$-.1373$
1.5	0	0	.0546	$-.1093$	$-.0169$	$-.0731$
2	0	0	$-.0023$	$-.0642$	$-.0275$	$-.0299$
3	0	0	$-.0166$	$-.0098$	$-.0124$	$+.0024$
4	0	0	$-.0055$	$+.0028$	$-.0015$	.0038
$+\infty$	0	0	0	0	0	0

All electric field components are normalized to the value of ReE_y at $y=-\infty$. The magnetic field parameters at the ocean floor H_x ($y>0$, $z=+0$) are normalized to the constant surface field H_x ($z=-0$). Note that for $z=-0$ one has $E_z=0$ when $y>0$. Coordinate y is expressed in skin-depths. These data are for the quasi-static and non-magnetic limit, and were derived from BAILEY (1977) and NICOLL and WEAVER (1977).

values of ReE_x (E-polarization) or ReE_y (H-polarization) at $z=0$, $y=-\infty$. Magnetic fields are normalized in similar fashion to ReH_y and ReH_x, respectively. The fields H_y and H_x at ($z=0$, $y=-\infty$) are taken as phase references and therefore chosen to be purely real.

The true complex ratios of E_x to H_y (E-polarization, Table 1) and of $-E_y$ to H_x (H-polarization, Table 2) are in fact given by the surface impedances Z_{xy} and $-Z_{yx}$. At ($z=0$, $y=-\infty$) these ratios are identical,

$$Z_{xy}(-\infty)=-Z_{yx}(-\infty)=(1+\mathrm{i})\left(\frac{\omega\mu_0}{2\sigma}\right)^{1/2}. \tag{2}$$

In the present geometry and coordinate system it is easy to show (see, e.g., FISCHER, 1975) that

$$Z_{xx}=Z_{yy}=0 \tag{3}$$

throughout the structure. Note also that for H-polarization, with the normalization convention and in the limit stated, $H_x=1$ at the surfaces of both land and ocean ($z=-0$), and is therefore not tabulated for $z=-0$. Also not tabulated is $E_z(y<0)$ at $z=-0$, since this field will be generated by surface charges that build up at the surface. The currents required to build and carry away these surface charges are small, i.e. of order $\omega\varepsilon/\sigma$ (c.f. PRICE, 1973), but the highly singular nature of our model ocean edge may result in significant vertical fields near the coast, to satisfy div E=0. At ($y>0$, $z=-0$), $E_z\equiv0$ however.

In Fig. 2 we give a representation of the surface impedance in terms of their real

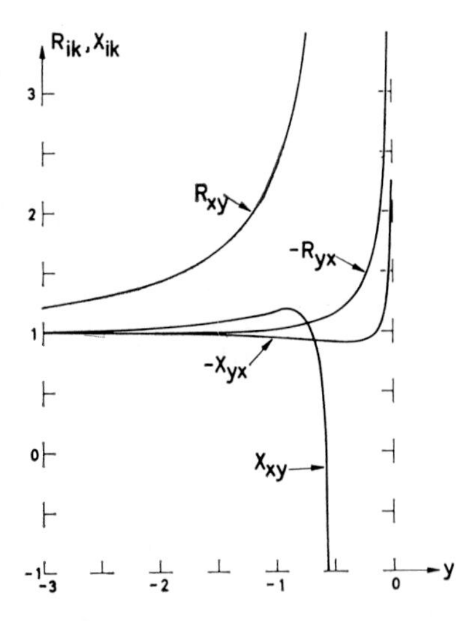

Fig. 2. Real and imaginary parts of the surface impedances for the structural model of Fig. 1. $Z_{xy}=R_{xy}+\mathrm{i}X_{xy}$ refers to E-polarization and $Z_{yx}=R_{yx}+\mathrm{i}X_{yx}$ to H-polarization induction. Units of y coordinate are skin-depths, while those of R_{ik} and X_{ik} are defined by Eq. (2). Close to the origin R_{xy} and X_{xy} rather suddenly, though continuously, vanish (this is not represented), whereas both R_{yx} and X_{yx} tend toward $-\infty$. For $y>0$, Z_{xy} and Z_{yx} both vanish, and in this geometry and coordinate system Z_{xx} and Z_{yy} are nil throughout the structure.

and imaginary parts,

$$Z_{ik} = R_{ik} + iX_{ik} , \tag{4}$$

where the ($i\ k$) subscript pair stands for (xy) or (yx). The most striking feature of Fig. 2 is the much larger distance to which the ocean coast effect is carried into the land for E-polarization than for H-polarization induction. As shown in more detail elsewhere (FISCHER *et al.*, 1978), this arises because the integrated ocean current I_x (current per unit length of ocean sheet, given by $I_x = H_y\,(z=-0) - H_y\,(z=+0)$) increases to very large values as the seashore is approached. In our idealized ocean model (Fig. 1) this integrated current even goes to infinity at the ocean edge ($y=+0$, $z=0$), where it acts as a very long line-antenna radiating the coast effect far afield. Under H-polarization induction the total integrated current (from $z=0$ to $z=+\infty$) is a constant, for reasons of continuity, with the consequent constancy of the surface magnetic field H_x as said above. Here the integrated ocean current $I_y = H_x\,(z=+0) - H_x\,(z=-0)$ increases monotonically from zero at the seashore to this total constant value as $y \to +\infty$. There is no line-antenna feature, and the coast effect is not carried to very large distances under H-polarization induction. Similar differences for the ranges to which perturbations are carried in the two polarizations can be expected for other two-dimensional structures, whenever large conductivity contrasts are encountered at shallow depths.

This work was supported in part by a grant from the Swiss National Science Foundation.

REFERENCES

BAILEY, R. C., Electromagnetic induction over the edge of a perfectly conducting ocean: the H-polarization case, *Geophys. J. R. Astr. Soc.*, **48**, 385–392, 1977.

FISCHER, G., Symmetry properties of the surface impedance tensor for structures with a vertical plane of symmetry, *Geophysics*, **40**, 1046–1050, 1975.

FISCHER, G., P.-A. SCHNEGG, and K. D. USADEL, Electromagnetic response of an ocean coast model to E-polarization induction, *Geophys. J. R. Astr. Soc.*, **52**, 599–616, 1978.

NICOLL, M. A. and J. T. WEAVER, H-polarization induction over an ocean edge coupled to the mantle by a conducting crust, *Geophys. J. R. Astr. Soc.*, **49**, 427–441, 1977.

PRICE, A. T., The theory of geomagnetic induction, *Phys. Earth Planet. Inter.*, **7**, 227–233, 1973.

Diakoptic Solution of Induction Problems

C. R. BREWITT-TAYLOR and P. B. JOHNS

Department of Electrical Engineering, Nottingham University, Nottingham, U. K.

(Received November 3, 1977; Revised January 17, 1978)

Two-dimensional induction problems are solved by means of an equivalent electrical network, which is then solved by the method of diakoptics. The network is divided into several subnetworks, which are partially solved independently of each other, and then are re-assembled into the full network. The advantages are that the size of arrays handled is reduced, and that it is possible to solve a problem which has most of the subnetworks the same as a previous problem more quickly than by solution without subdivision, since the common subnetworks need not be re-solved. A computer program has been written to use this method, and an example is given with storage and time used.

1. Introduction

The problem to be considered is the usual two-dimensional induction problem, in which fields and conductivities vary only in the y and z directions, and the fields vary with an assumed exp $(i\omega t)$ time variation. In this case Maxwell's equations reduce to the well-known diffusion type of equations for the two separate polarisation cases. In E-polarisation:

$$\frac{\partial^2 E_x}{\partial y^2}+\frac{\partial^2 E_x}{\partial z^2}=i\omega\mu_0\sigma E_x\,, \tag{1}$$

and in H-polarisation:

$$\frac{\partial}{\partial y}\left(\frac{1}{\sigma}\frac{\partial B_x}{\partial y}\right)+\frac{\partial}{\partial z}\left(\frac{1}{\sigma}\frac{\partial B_x}{\partial z}\right)=i\omega\mu_0 B_x\,. \tag{2}$$

Here all the symbols have their usual meanings. In these equations we are making the usual geophysical assumptions of negligible displacement currents and non-magnetic materials, though the method described in this paper can be simply extended to remove these assumptions.

It is well known that there is an analogy between Maxwell's equations in one or two dimensions and the equations relating voltage and current in a transmission line or surface; and similarly that a five-point difference approximation to the differential equations above has the same form as the equation describing an electrical mesh network of the kind shown in Fig. 1. This is discussed in a geophysical context by SWIFT (1971). In the present work the components of the network are considered to outline rectangles in each of which the conductivity is constant, and the difference equations represented are those of BREWITT-TAYLOR and WEAVER (1976). In the E-polarisation case the voltages

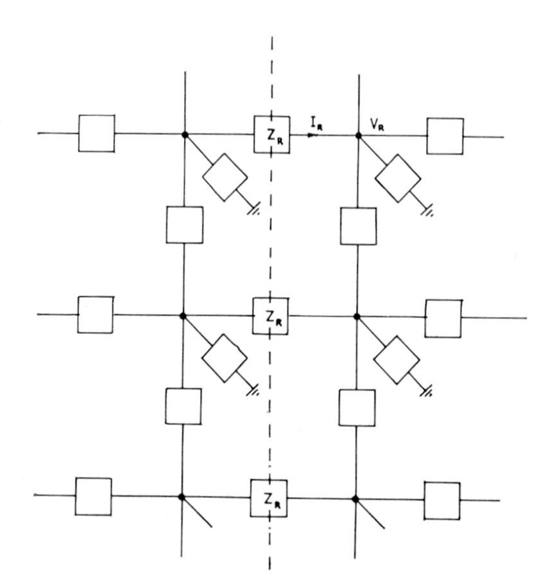

Fig 1. Part of the network representing the differential equation. The dashed line shows part of a diakoptic subnetwork boundary.

at the junctions of the network represent the electric field E_x; for a regular mesh of distance h between junctions, the transverse components are inductors of value μ_0 Henry, and the components to ground are resistors of value $1/h^2\sigma$ Ohms. In the H-polarisation case the junction voltages represent the magnetic field B_x; the transverse components are resistors of value σ Ohms, and the components to ground are capacitors of value $1/h^2\mu_0$ Farads. In the case of uneven mesh spacings and conductivity boundaries, suitably averaged values of conductivity are used. However the diakoptic method described here does not depend on the values of the components, but only on the structure of the network, and so it may be used for either polarisation case. Both are special cases of the network in Fig. 1. In this network the voltages at adjacent junctions obey the following equation, which follows from Ohm's law and the conservation of current:

$$\frac{(V_1-V_0)}{Z_1}+\frac{(V_2-V_0)}{Z_2}+\frac{(V_3-V_0)}{Z_3}+\frac{(V_4-V_0)}{Z_4}+\frac{V_0}{Z_0}=0\,, \tag{3}$$

where V_1 to V_4 are the voltages at the junctions surrounding a typical junction with voltage V_0, Z_1 to Z_4 are the corresponding transverse impedances and Z_0 is the impedance from the central point to ground.

Diakoptics is a general method for the solution of large electrical networks, originated by KRON (1963); a useful introduction is given by BRAMELLER *et al.* (1969). The method proceeds by dividing up the network into a number of subnetworks, which are analysed separately. For each subnetwork we analyse its internal properties, and then set up a matrix which describes it as it is seen from the outside along the branches connecting it to other subnetworks. The circuit is then re-assembled using these matrices to solve for the currents in the connecting components. One potential advantage is that we avoid handling the entire problem at once, and so reduce the size of arrays in a computer program from problem-sized to subnetwork-sized. Another advantage is that subnetwork results can be kept and used again without re-calculation in other problems containing the same subnetworks; this allows us to solve several similar problems more quickly than by complete re-solution for each problem.

2. Diakoptic Solution

The network is divided into subnetworks by boundary lines parallel to the coordinate axes between the lines of junctions (Fig. 1); the boundary lines need not run straight through the whole network. Every junction is within one subnetwork, but some of the circuit branches cross the boundary lines. We introduce as additional unknowns the currents I_R in all these branches; these components are then taken out of the network, leaving a number of separate subnetworks. The removed branch currents I_R appear as source currents for the subnetworks.

The first stage of computation is to analyse the internal behaviour of each section. For each junction within the subnetwork we have an equation of the form of Eq. (3). Some terms in this general form may be missing if the junction is next to a diakoptic boundary, or at the edge of the problem where the equation is modified to account for the boundary conditions. There may also be impressed source currents appearing on the right-hand side. All these equations are gathered into matrix form, with a suitable ordering of the junctions:

$$\mathscr{C} \cdot \boldsymbol{V} = \boldsymbol{I}_0 + \boldsymbol{I}_R \,. \tag{4}$$

The matrix $\mathscr{C}$ contains all the equation coefficients (obtained from the circuit component values); it is symmetric and very sparse, having not more than five non-zero elements in each row or column. The vector $\boldsymbol{V}$ consists of all the unknown junction voltages. The right-hand side consists of current sources I_0 arising from boundary conditions at the edge of the problem, and also the currents I_R in removed branches adjoining this subnetwork. In our computer program it is assumed that $\boldsymbol{I}_0$ is known at this stage, so that boundary conditions must be specified at the start. The currents $\boldsymbol{I}_R$ are not yet known, but the matrix $\mathscr{C}$ is completely determined.

Next we compute the inverse of $\mathscr{C}$. We have used a special method designed for sparse matrices, originated by Zollenkopf (1971). This method stores only the non-zero elements of the matrix, with pointer arrays to locate them; it also obtains the inverse in a factored form which reduces the number of non-zero coefficients that must be stored.

The second stage is to construct a matrix $\mathscr{S}$ which describes the subnetwork from outside. Viewing the subnetwork as a black-box circuit with external terminals at the junctions where removed branches were attached, the matrix $\mathscr{S}$ relates the currents I_R and the voltages V_R at these terminals:

$$\boldsymbol{V}_R = \boldsymbol{V}_0 + \mathscr{S} \cdot \boldsymbol{I}_R \,. \tag{5}$$

$\boldsymbol{V}_0$ is the vector of terminal voltages when none of the removed branches is carrying any current; they arise from sources representing boundary conditions at those edges of the problem included in this subnetwork. Each column of $\mathscr{S}$ consists of the terminal voltages when just one of the terminal currents is unity, and the rest are zero, and the problem boundary sources are removed. These terminal voltages are obtained using the inverse matrix $\mathscr{C}^{-1}$; from Eq. (4) we have:

$$\boldsymbol{V}=\mathscr{C}^{-1}\cdot\boldsymbol{I}_0+\mathscr{C}^{-1}\cdot\boldsymbol{I}_R\,. \tag{6}$$

Comparing Eq. (5) and (6) we can identify $\boldsymbol{V}_0$ with $\mathscr{C}^{-1}\cdot\boldsymbol{I}_0$. Also each column of $\mathscr{S}$ is then $\mathscr{C}^{-1}\cdot\boldsymbol{I}_R$, where $\boldsymbol{I}_R$ has one unit element and the rest zero. We carry out a series of such matrix multiplications making each element of $\boldsymbol{I}_R$ unity in turn, to obtain all the columns of $\mathscr{S}$. In all these matrix multiplications we throw away the voltages at the internal junctions of the subnetwork, which result from the calculations at the same time as the terminal voltages. Because $\mathscr{C}^{-1}$ is only held in a factored form, actual matrix multiplications must be performed here, rather than just selecting terms out of an explicitly known matrix, which the multiplications are equivalent to.

Computations on the individual subnetwork are now complete. We save for future use the following information about the subnetwork: (a) the removed branch impedances Z_R, (b) the black-box matrix $\mathscr{S}$, (c) the terminal voltages $\boldsymbol{V}_0$, (d) a record of problem boundary sources within the subnetwork, and (e) the inverse matrix $\mathscr{C}^{-1}$. Each other subnetwork is then analysed similarly.

The final stage is to re-assemble the network. We set up a network consisting of the black-box circuits of each subnetwork, joined together by the removed components. Apply Ohm's law to each removed component:

$$I_R=(V_{R1}-V_{R2})/Z_R\,, \tag{7}$$

where V_{R1} and V_{R2} are the terminal voltages at the ends of this component. Then we eliminate the voltages from this equation, using Eq. (5). The result is a set of linear equations relating the removed branches currents I_R to each other, which we write schematically in the form:

$$\boldsymbol{I}_R=\{(\boldsymbol{V}_{01}+\mathscr{S}_1\cdot\boldsymbol{I}_{R1})-(\boldsymbol{V}_{02}+\mathscr{S}_2\cdot\boldsymbol{I}_{R2})\}/Z_R\,. \tag{8}$$

Each current is related to all other currents that border the two subnetworks it connects. In practical cases the matrix of coefficients of these equations is reasonably full, and we solve it by a standard elimination method for full matrices. Considerable effort is needed in setting up these equations, and elsewhere, to keep track of the connections of the various removed components and to be consistent in sign conventions for the currents.

Having solved these equations and obtained all the removed branch currents, we can select those that are relevant to a particular subnetwork, and substitute them in Eq. (6) to obtain all the junction voltages. When this is done for all subnetworks, the solution of the problem is complete.

3. Numerical Example

A computer program in FORTRAN has been written to use the diakoptic method. As an example we chose a simple geophysical model that one of us has used as an example in a previous paper (BREWITT-TAYLOR and WEAVER, 1976). Since diakoptics is a new method of solving the numerical equations only, and not a new method of setting them up, the solutions are exactly the same as those obtained in the previous paper, and so are not shown again here. The problem is shown in Fig. 2. It was solved for the E-polari-

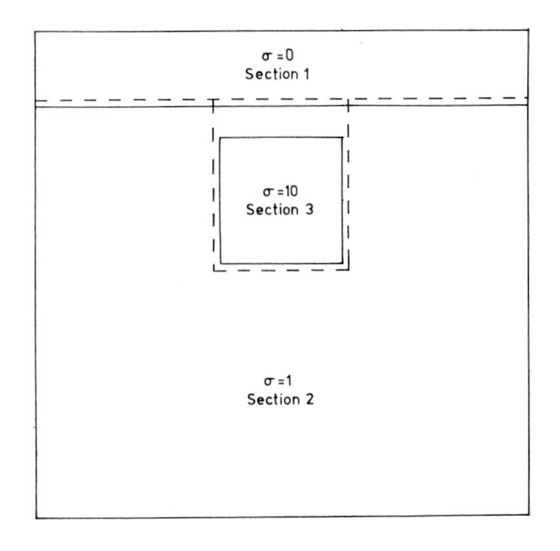

Fig. 2. The example problem. The solid lines show conductivity boundaries, the dashed lines show subnetwork boundaries.

sation case. The problem was first solved without division into subnetworks, using the same sparse-matrix inversion technique mentioned above; this is easily achieved in the program by omitting the steps relating to removed branches. It was then solved again by division into three subnetworks, as shown by the dashed lines in Fig. 2. The first subnetwork, representing the atmosphere, would only need solving once for any *E*-polarisation problem with this frequency and horizontal mesh spacing. The second subnetwork, representing a uniform region around the anomalous region, can be used for many anomalies. These are then obvious candidates for saving their subnetwork results for repeated use. The anomalous conductivity is confined within the third subnetwork, which will need to be re-computed for each anomaly used. In seeking to match an observed surface field distribution, it will be useful to try many such anomalous conductivities. With this subdivision one can use any anomaly that can be represented in the grid mesh in the third subnetwork, not necessarily the uniform block used here.

Computer times and storage estimates are shown in Table 1. The storage used depends greatly on how much programming effort is put into use of disc storage to save core storage. Subdivision gives a storage saving, as we avoid handling the very large inverse matrix for the whole problem. Looking at the computer times, we see that there is a considerable overhead in the extra computation of the diakoptic method, particularly

Table 1. Computer time and storage for the example problem.

Operation	Computer time	Storage
Section 1 (air)	14 sec	14 K
Section 2 (surrounds)	134	59 K
Section 3 (anomalous)	7	8 K
Joining sections	13	27 K
Total for first problem	168	59 K
Subsequent problems	20	27 K
Undivided solution	71	78 K

Storage is in units requiring one for an integer and two for a real number; the figures do not include storage for arrays that are independent of the subdivision (10 K in this case).

in analysing the two large subnetworks. Much of this is spent doing the many matrix multiplications involved in obtaining the matrix $\mathscr{S}$. Thus for a single problem it is quicker to solve it without subdivision. To change the anomalous conductivity, however, we only need to analyse the small third subnetwork and re-assemble the problem, which in this example takes less than a third of the time of a complete re-solution. Thus we only need to solve a few anomalous conductivities before we have made a time saving as compared with complete undivided re-solution for each problem. The time taken for subnetwork assembly depends largely on the total number of removed branches; so for a quick solution of repeated problems it is desirable to minimise this number, and also the size of the subnetwork containing the anomaly. A compromise has to be reached between this consideration and the large time and storage required by large non-anomalous subnetworks and the flexibility provided by the subnetwork structure. Thus for this type of problem it would be quickest to use only two subnetworks (the anomalous region, and everywhere else), but this has the disadvantage that the normal subnetwork would need re-analysing for every change of surrounding medium, which would be a long process.

4. Conclusion

The diakoptic method provides a computer storage saving in any induction problem, at the expense of extra computation time. In the case of several problems in which only a small area of conductivity changes from one problem to the next, such as arise naturally in the interpretation of geophysical data, the method gives solutions to problems after the first considerably more quickly than repeated solution without subdivision. The solutions obtained are exactly the same as solutions of the same difference equations by more direct methods.

There is no need, of course, to express the problem in terms of an electrical network and the method of diakoptics or substructures may certainly be applied to the simultaneous equations assembled by the differencing process. However, in forming the substructures there is much more meaning in cutting a graph which corresponds directly to the physical problem rather than cutting a matrix which relates to the physical problem in a much less direct way. Thus there is considerable advantage in forming a graph. The use of electrical terms to describe the graph is merely a matter of convenience of language for the authors.

This work was carried out as part of a research contract from the U. K. Ministry of Defence.

REFERENCES

BRAMELLER, A., M. N. JOHN, and M. R. SCOTT, *Practical Diakoptics for Electrical Networks*, 242 pp., Chapman and Hall, London, 1969.

BREWITT-TAYLOR, C. R. and J. T. WEAVER, On the finite difference solution of two-dimensional induction problems, *Geophys. J. R. Astr. Soc.*, **47**, 375–396, 1976.

KRON, G., *Diakoptics*, MacDonald, London, 1963.

SWIFT, C. M., Theoretical magnetotelluric and turam response from two-dimensional inhomogeneities, *Geophysics*, **36**, 38–52, 1971.

ZOLLENKOPF, K., Bi-factorization—Basic computational algorithm and programming techniques, in *Large Sparse Sets of Linear Equations*, edited by J. K. Reid, pp. 75–96, Academic Press, London, 1971.

Induction by *Sq*

W. D. PARKINSON

Geology Department, University of Tasmania, Hobart, Australia

(Received November 8, 1977; Revised February 12, 1978)

The earliest estimates of the conductivity of the earth were based on the analysis of *Sq*. The picture that emerged from analyses performed up to 1940 was that of a shallow conducting layer at the surface underlain by a few hundred kms in which the conductivity is low followed by a region of very rapid or discontinuous increase of conductivity. The overall pattern now considered most likely is not essentially different, but is now realised that the period range spanned by *Sq* is too small. Furthermore, when trying to match a radial model of conductivity, *Sq* data is biased by conductivity anomalies. Longer period variations seem to be less subject to surface conductivity irregularities.

Important progress has been made towards estimating the conductivity structure beneath oceans; it seems to be very different from that under continents. The work of Larsen on Oahu indicates a distribution in which the conductivity is an order of magnitude greater for the top 300 km. This, and similar work off the coast of California, have done much to elucidate the behaviour of the geomagnetic field at the edge of oceans.

Sq periods have proved valuable in investigating conductivity anomalies also explored by shorter period variations. Generally anomalies that influence the field at sub-storm periods also appear at diurnal frequencies, but "dead-*Z*" anomalies of the Tuscon type often do not.

1. Introduction

The subject of electromagnetic induction in the earth may be said to have started with Schuster's investigations into the origin of the solar diurnal variation, *Sq*, in 1887. The horizontal components of the *Sq* field could be generated either by two overhead current loops, anticlockwise (in the northern hemisphere) and clockwise (in the southern hemisphere) or by two underground loops in the opposite directions. But the vertical components would be different for these two current systems, and it was found that the observed sense of the vertical component is that caused by overhead currents not underground currents.

Two years later SCHUSTER (1889) looked at the problem more quantitatively. He found that the vertical component is smaller than it should be if the *Sq* field is generated entirely by overhead currents. He assumed that the inner part was due to induction and remarked "This we might have expected". He was able to show that a conductivity much higher than that of surface rocks was required to explain his data, and he introduced a model in which there is a non-conducting shell about 1,000 km thick overlying a good but finitely conducting "core". This model has changed only in detail since that time.

Over the years a great deal of work has gone into interpreting the ratio of internal to external parts of the *Sq* potential in terms of a radial conductivity distribution in the earth. Highlights have been the work of CHAPMAN (1919), CHAPMAN and WHITEHEAD (1922), CHAPMAN and PRICE (1930), and LAHIRI and PRICE (1939). This has been well summarised several times (e.g. PRICE, 1967).

2. Global Conductivity Distribution

The conventional method of analysis of *Sq* data is to derive Fourier coefficients for each element at each observatory for periods 24 hr, 12 hr etc. and express each of these as a function of latitude and longitude by spherical harmonic analysis. It might be thought that the harmonic coefficients so derived would be among the most reliable of geomagnetic data. Power density spectra show that *Sq* is the most prominent variation in a wide frequency spectrum and good quality data is available from a fairly extensive net-work of observatories. But it is becoming apparent that the data are unreliable for global studies. Several workers, such as SCHMUCKER and JANKOWSKI (1972) and JADY (1974a) have pointed out that *Sq* is quite strongly affected by conductivity anomalies and by the oceans. Also the distribution of observatories makes it difficult to allow for the influence of the equatorial electrojet. SUZUKI (1973) showed that the principal harmonics depend on how the electrojet is allowed for.

In spite of the comparatively large proportion of the total variation power contained in *Sq* the harmonic coefficients are not very precisely known. A wide variation is found in the ratio and phase difference between the external and internal potential coefficients of the principal *Sq* terms (PARKINSON, 1974).

JADY (1974b) introduced a model in which a finitely conducting shell overlies a superconducting core. His average result indicates a resistivity of about 50 ohm-m in a 600 km thick layer.

On the other hand *Dst* data, derived from magnetic storm recovery fields is geographically more consistent. GRAFE (1963) found similar response functions for Apia, Kakioka, and Furstenfeldbruck (island, coastal, and inland sites).

Since about 1970 the emphasis has changed to an analysis of longer period variations. The frequency spectrum covered by reliably determined *Sq* harmonics extends only from 1 to 3 cycles per day. A wider frequency range is necessary if anything better than a rough indication of global conductivity is to be obtained.

The leading work in analysing the extended frequency spectrum of variations and applying it to global conductivity studies has been that of BANKS (1972). He considers that the role of *Sq* is subordinate, and that it is better to determine a distribution of conductivity from storm time variations and modify it at the shallow end to agree with *Sq*.

At the same time there has been an increase in the sophistication of the methods used to invert the data. BAILEY (1973) suggested a method based on the Kramers-Kroenig relation, but it requires unrealistically complete data. The technique of Backus and Gilbert as used by PARKER (1970) seems to be more practical, but the iterative process sometimes fails to converge. This is particularly true if some of the data is contaminated by conductivity anomalies, so that it does not apply to the model being used in the inversion process.

Another interesting inversion method was introduced by SCHMUCKER (1970a) in his classic study of induction anomalies in California. A complex parameter C, called the inductive scale length can be obtained for each frequency from either the ratio of Z/H or E/H (Z, E are the vertical magnetic and horizontal electrical fields, and H represents some horizontal component of the magnetic field). The real part of C indicates the average depth at which the eddy currents are flowing and the imaginary part is a "skin depth" simply related to the resistivity of the medium in which the eddy currents are flowing. The difficulty in using the ratio Z/H is that the wave number of the primary field must be known.

3. Anomalous Observatory Data

The most likely reason for the inconsistencies in *Sq* is that certain observatories are affected by conductivity anomalies. So it might be thought that useful information about anomalous parts of the earth could be obtained from observatory data. Not much has been achieved in this direction, largely because of the wide spacing of observatories. A start has been made by Berdichevsky and his collaborators (BERDICHEVSKY *et al.*, 1976), and by Schmucker. When Fourier coefficients are analysed into spherical harmonics the minimum number of observatories necessary to perform the analysis is equal to the number of harmonics used in the representation of the Fourier coefficients. If each pair of Fourier coefficients can be represented by just one harmonic, then just one observatory suffices to describe the global distribution. Fortunately one harmonic is predominant for each sine wave, for instance the P_3^2 harmonic for the 12 hr sine wave. SCHMUCKER's (1970b) technique is to determine a harmonic coefficient for each observatory. Because the phase, as well as the amplitude of the sine waves representing the components of *Sq*, is available, a complex "inductive scale length" can be obtained. The results obtained by Schmucker for the 12 and 8 hr waves form a cluster with a spread of a factor of 2.5 in depth and more than a factor of 10 in conductivity. It is typical of all induction studies to obtain less precision in conductivity than in depth. This is because the conductivity can be greatly influenced by a small change in the phase of the complex coefficients. The scatter probably partly reflects the influence of the neglected harmonics. However, the observatory with the deepest effective depth is Wingst which is known to lie close to a lateral inhomogeneity in conductivity.

The complex scale length can be determined from either the ratio of Z/X or Z/Y. In theory these should agree, in other words there should be a predictable relation between the north and east components, depending only on latitude. In particular they should be in time quadrature. Two orthogonal horizontal directions can be found such that the components in those directions are in quadrature, and these can be defined as the effective north and east. Then a virtual latitude can be found which agrees with the ratio of the virtual north and east components. This is the latitude used to determine the scale length.

If the Fourier coefficients could be exactly described by the spherical harmonic of degree $(m+1)$ there would be a linear relation between each Fourier coefficient and the appropriate function of latitude. For instance the Fourier coefficients of the east component would satisfy

$$A_m = C_A P_{m+1}^m(\theta)/\sin\theta$$
$$B_m = C_B P_{m+1}^m(\theta)/\sin\theta$$

where C_A, C_B are independent of latitude, θ is geographic colatitude. Figure 1 shows a plot of the Fourier coefficients for $m=2$ for a number of observatories. The departure from a straight line passing through the origin is considerable and systematic, suggesting that other harmonics are important. If the contribution of other harmonics is subtracted from the Fourier coefficients they conform more closely to the linear relation. Figure 2 shows a similar plot of the same Fourier coefficients reduced by the spherical harmonic coefficients of degree other than $n=m+1$. The amount of reduction is derived from a global analysis (PARKINSON, 1971).

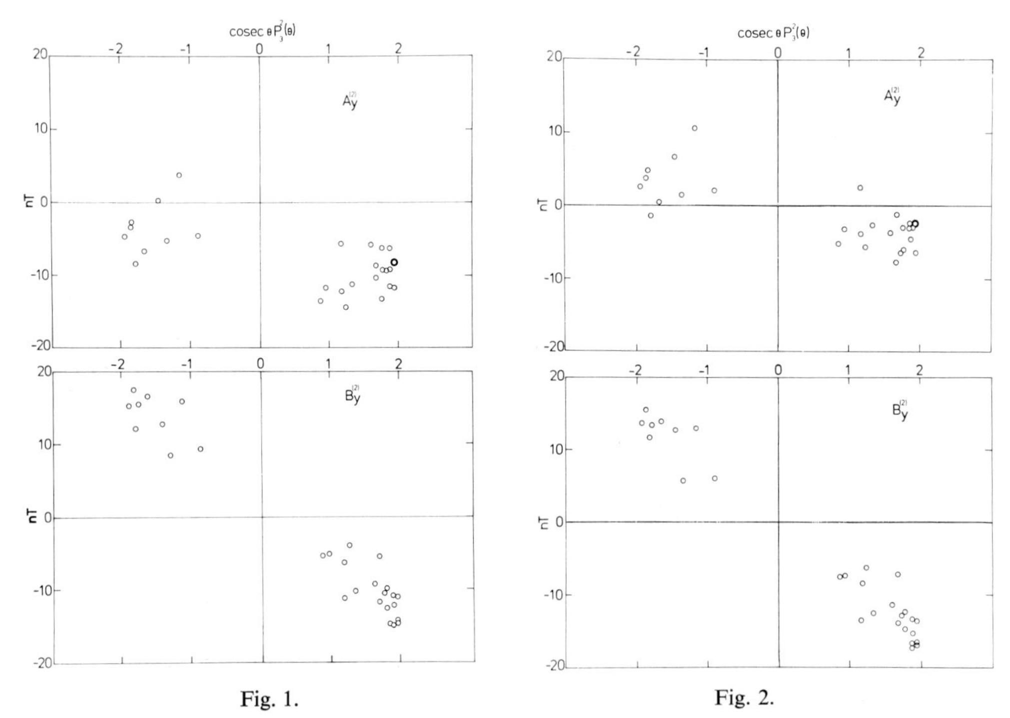

Fig. 1. Observed Fourier coefficients for 12-hr period of east component plotted against cosec $\theta \cdot P_3^2(\theta)$.

Fig. 2. Fourier coefficients of Fig. 1 reduced by subtracting the contributions of terms other than $n=m+1$ plotted against cosec $\theta \cdot P_3^2(\theta)$; θ is geographic colatitude.

From each pair of reduced Fourier coefficients a value of the complex inductive scale length has been derived. The results are shown in Fig. 3. If the conductivity structure is as suggested by Schuster, all points should lie in the half quadrant with phase angle between $-45°$ and $0°$. Most points do lie in this region and suggest a depth of

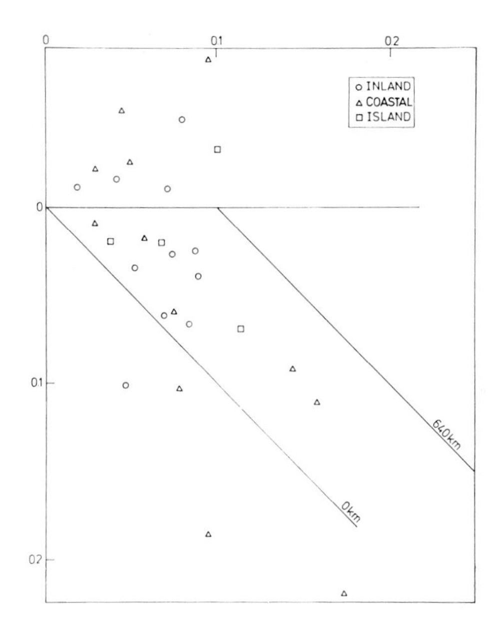

Fig. 3. Inductive scale-length for individual observatories plotted in the complex plane. Ordinate and abscissa are in units of earth radius. Assuming a Schuster type model the sloping lines represent depths to the top of the conductor. Fourier coefficients plotted in Fig. 2 have been used in the computation.

eddy currents between 200 and 600 km. Four points have phase angles between $-90°$ and $-45°$ suggesting a conductivity decreasing with depth. Three of these are coastal sites, but several other coastal sites plot in different regions of the diagram. A number of inductive scale lengths have positive phase angles, which is physically impossible for a layered conductivity structure.

Although several of the more erratic points represent islands or coastal sites there is not enough correlation with site location to suggest this will be a useful way of searching for conductivity anomalies.

BERDICHEVSKY *et al.* (1976) have been able to draw limited contours of depth to the conductosphere for regions in which there is a dense network of observatories. Their method involves making a flat earth approximation, instead of using only one harmonic, and using the 6 hr period sine wave.

Both Schmucker's and Berdichevsky's techniques depend on the conductivity being a function only of depth. At a site near a lateral inhomogeneity the vertical component will be abnormally large. For instance the calculations of ASHOUR (1973) and of HOBBS (1975) for a hemispherical ocean show a distinct increase in Z near the coast. This increases the amplitude of C and therefore gives a spuriously large depth and skin depth, i.e. a spuriously low conductivity. This effect might well apply to a number of observatories that are included in the world average. Thus global estimates of conductivity structure are likely to be influenced by a systematic error which will bias them towards too great a depth and too low a conductivity. This would mean that the iso-conductivity surfaces would tend to be depressed in estimates based on global *Sq* analyses. This appears to happen for 12 hr but not for 24 hr.

RIDDIHOUGH (1967) studied correlations between Z and H at observatories in the British Isles, France and Belgium. He concluded that induction in the Atlantic Ocean, as predicted by RODEN (1964) controls contours of amplitude and phase. This was before the anomalous nature of Eskdalemuir was known, and this may have some influence on Riddihough's results.

RIKITAKE (1959) found that the phase of Z was retarded at Aso relative to other Japanese observatories. This infers that the eddy currents are flowing in a poorer conductor near Aso.

A very simple data treatment is to consider the daily ranges of the three elements H, D, and Z as a function of position. Abnormal Z ranges have been found in the seismic regions of India by SRIVASTAVA *et al.* (1974). Unfortunately these were based on Z/H ratios rather than Z/D ratios although the grain of the sub-continent is more or less N-S. A similar survey based on daily ranges in F (total intensity) along the Atlantic coast of Canada by EDWARDS and GREENHOUSE (1975) suggests a high conductivity inland rather than control by the ocean.

Daily ranges have also been used by FANSELAU (1968) to delineate the north German anomaly.

4. *Sq* Surveys for Conductivity Anomalies

A number of surveys using arrays of temporary magnetographs have been designed specifically to investigate conductivity anomalies at d.v. (diurnal variation) frequencies. This has the advantage over observatory data of a much closer spacing of observing sites, but the disadvantage that a long sequence of daily records to determine harmonic coefficients is lacking. However, LILLEY (1975a) has suggested that there may be advantages in investigating individual days.

Bondarenko and Kistas (ADAM, 1976) analysed an extensive survey of the Carpathian region in which they plotted a parameter depending on the ratio of combined horizontal components of *Sq* to the vertical component, relative to a reference site (Lvov observatory). The high heat flow regions of the east Carpathians are characterised by high values of this parameter.

Another large array survey covered N.W. USA and S.W. Canada. Shorter period variations had been found to show an attenuation of the vertical component west of the Plains-Cordillera boundary (CANER. 1970). However, the d.v. vertical component power at sites east and west of this boundary (cf. Penticton and Pincher) is similar (CANER, 1971), showing that the attenuation of the vertical component in this area does not extend to d.v. frequencies. The extreme attenuation of short period Z variations at Tucson also does not extend to d.v. frequencies. This dissimilarity between short period and d.v. anomalies suggests that they may be looking at quite different depths. This was suggested by GARLAND (1971).

The extended anomaly which passes near the Black Hills of South Dakota is evident at periods up to 24 hr. However, the geographic pattern changes abruptly between 12 and 24 hr periods. CAMFIELD and GOUGH (1975) ascribe the 24 hr effect to the varying nature of the medium through which the magnetic field from a distant source must pass. The Basin and Range Province is known to be a region of high conductivity. The mechanism envisaged is that of a secondary field induced by *Sq* in the Basin and Range

Province. This in turn induces currents in conductivity anomalies in the Yellowstone plume area which give rise to what might be called a "tertiary" field. Such an anomaly is called a VARTRAN anomaly.

Some of the most definitive work with arrays at d.v. frequencies has been done by Lilley and his co-workers (LILLEY, 1975b). BENNETT and LILLEY (1973) analysed Fourier coefficients of 24 to 6-hr period over an extensive synoptic network in S.E. Australia. In a number of contour maps of amplitude and phase the dominant feature is rapid change in amplitude and phase of Z on going inland from the east coast. This will be discussed in more detail below. Another feature is the enhanced H and D over the Otway anomaly in the S.W. of the survey area. The vertical component seems to be abnormally low over the whole of southern N.S.W., Victoria, and northern Tasmania. However, this does not extend to southern Tasmania (BISDEE, 1977). It is interesting that the shallow water of Bass Strait seems to have no channelling effect at periods as long as 6 hr.

5. *Sq* and the Ocean

Much has been written on the influence of the oceans on *Sq*. Speculation dates from the time of CHAPMAN and WHITEHEAD (1922). Most successful models of global conductivity require a thin conducting layer at the surface. JADY (1974a) showed that this has a maximum conductance equivalent to an ocean 1 km deep.

There are two pressing reasons for wanting an accurate evaluation of the secondary field from induction in ocean water. One is to improve the estimate of global conductivity by correcting each site for the presence of oceanic conductivity. The other is to elucidate the obvious connection between anomalies in *Sq* and the ocean.

The first attempt to make a quantitative estimate of the effect of ocean-water seems to be that of DE WET (1949). In his calculations self-inductance was ignored. After a great deal of careful calculation RIKITAKE (1962) concluded that when self-induction and the effect of the conductosphere were taken into account, the influence of the actual oceans on *Sq* would be only minor. On the other hand RODEN (1964) calculated that a flat conducting strip simulating the Atlantic Ocean would have a profound effects on *Sq*. Rikitake's model was more realistic than that of Roden. HOBBS (1975) also calculated the eddy currents induced in the sea water by *Sq*, but did not translate them into a secondary field.

The most successful calculations of the secondary field from the oceans was that of BULLARD and PARKER (1970) based on a method originally devised by PRICE (1949). What Bullard and Parker actually produced was a map showing the current function for a 24 hr period sine wave, from which it would have been possible to derive the complete secondary field. RICHARDS (1970) continued these calculations. The only application of either of these results to actual situations appears to be that by BENNETT and LILLEY (1973), and LILLEY and PARKER (1976).

Bennett and Lilley operated an array of magnetometers in S.E. Australia mainly to investigate the coast effect in *Sq*. They estimated the effect of sea water from Richards calculations, and found it to be inadequate, and in the wrong phase, to explain the sudden phase variation in Z near the east coast. Similar results were found by Lilley and Parker for both the east and west coasts of Australia.

It appears, at least as far as the 24 hr wave is concerned, that the coast effect cannot be explained by induction in ocean water unless, at the same time, there is a distinctly shallower depth to the conductosphere under the oceans than under the continents. The conductivity distribution in suboceanic crust and mantle is becoming clearer, thanks mainly to the work of the Scripps group on the ocean floor and on island (see Larsen's contribution in this issue). This can be done in spite of logistic difficulties and the influence of tidal ocean currents. A magneto-telluric survey on Oahu by LARSEN (1975) covered a wide range of frequencies. The response function suggested a distribution of conductivity with depth quite different from that found on continents. The two distributions approach each other at depths of about 600 km, but the low conductivity zone, almost universally found on continents, appears to be missing under the oceans. Unfortunately the Hawaiian Islands may not be typical of oceanic crust and mantle, and confirmation of these results from elsewhere would be welcome.

6. Tidally Generated Diurnal Variations

The dynamo action of motion of a conductor relative to the earth's main field can take place in the oceans as well as in the ionosphere. HILL and MASON (1962) investigated the diurnal variation of total intensity in England and on a buoy at sea nearby. They found the variation nearly twice as great at sea and attributed this to the field produced by tidal flow in the ocean. More recent work by LARSEN and COX (1966) indicates that the lunar quiet day variation Lq is largely generated in the oceans.

MALIN and WINCH (1968) found that the lunar daily variations were anomalously large at several coastal observatories such as Hartland and Watheroo. Although Watheroo is 80 km from the coast there is evidence (EVERETT and HYNDMAN, 1967) that the coastal sedimentary basin has a high conductivity, and Watheroo, near the eastern edge of the basin may be near a "virtual" coastline.

In the careful study by LARSEN (1975) on Oahu the data were filtered to remove the diurnal variation and its harmonics to avoid contamination with tidally produced fields. It is interesting to see the d.v., which for many years was the only geomagnetic variation sufficiently well known to contribute to induction studies, now being purposely excluded.

REFERENCES

ADAM, A., *Geoelectric and Geothermal Studies*, 600 pp., Akademiai Kiado, Budapest, 1976.

ASHOUR, A., Theoretical models for electromagnetic induction, *P.E.P.I.*, **7**, 303–312, 1973.

BAILEY, R. C., Global geomagnetic sounding, *P.E.P.I.*, 7, 234–244, 1973.

BANKS, R. J., Overall conductivity distribution of the earth, *J. Geomag. Geoelectr.*, **24**, 337–351, 1972.

BENNETT, D. J. and F. E. M. LILLEY, An array study of daily magnetic variations in south-east Australia, *J. Geomag. Geoelectr.*, **25**, 39–62, 1973.

BERDICHEVSKY, M., E. B. FAINBERG, N. M. ROTANOVA, J. B. SOHIROV, and Y. B. VANYAN, Deep electromagnetic investigations, *Ann. Geophys.*, **32**, 143–156, 1976.

BISDEE, C., *Pers. Com.*, 1977.

BULLARD, E. and R. L. PARKER, Electromagnetic induction in the oceans, in *The Sea*, *4*, edited by A. E. Maxwell, pp. 695–730, John Wiley and Sons, New York, 1970.

CAMFIELD, A. and D. I. GOUGH, Anomalies on daily variation magnetic fields, *Geophys. J.*, **41**, 193–218, 1975.

CANER, B., Electrical conductivity structure in western Canada, *J. Geomag. Geoelectr.*, **22**, 113–127, 1970.

CANER, B., Quantitative interpretation of geomagnetic depth sounding data in western Canada, *J. Geophys. Res.*, **76**, 7202–7216, 1971.

CHAPMAN, S., The solar and lunar diurnal variations of terrestrial magnetism, *Phil. Trans. Roy. Soc. London*, **A218**, 1–118, 1919.

CHAPMAN, S. and A. T. PRICE, The electric and magnetic state of the interior of the earth as inferred from terrestrial magnetic variations, *Phil. Trans. Roy. Soc. London*, **A229**, 427–460, 1930.

CHAPMAN, S. and T. T. WHITEHEAD, The influence of electrically conducting material within the earth on various phenomena of terrestrial magnetism, *Trans. Phil. Soc. Cambridge*, **22**, 463–482, 1922.

DE WET, J. M., Numerical methods for solving electromagnetic induction problems, Thesis, University of London, 1949.

EDWARDS, R. N. and J. GREENHOUSE, Geomagnetic variations in eastern U.S.A., *Science*, **188**, 726, 1975.

EVERETT, J. E. and R. D. HYNDMAN, Geomagnetic variations and electrical conductivity structure in south-west Australia, *P.E.P.I.*, **1**, 24–34, 1967.

FANSELAU, G., The use of range differences for the interpretation of conductivity anomalies, *P.E.P.I.*, **1**, 177–180, 1968.

GARLAND, G. D., Electrical conductivity anomalies—mantle or crust?, *Comments on Earth Science: Geophysics*, **1**, 167–172, 1971.

GRAFE, A., Die Bedeutung der Abweichungen geomagnetischer Tagesmittel vom sogenannten Normalwert für die Analyse des geomagnetischen Ringstrom-effektes, *Deutsch Akad. Wiss. Berlin, Geomagn. Inst. Potsdam, Abh.*, **31**, 7–53, 1963.

HILL, M. N. and C. S. MASON, Diurnal variation of the Earth's magnetic field at sea, *Nature*, **195**, 365–366. 1962.

HOBBS, B. A., Electromagnetic induction in the oceans, *Geophys. J.*, **42**, 307–313, 1975.

JADY, R. J., Conductivity of earth models, *Geophys. J.*, **36**, 399–410, 1974a.

JADY, R. J., Conductivity models and Banks data, *Geophys J.*, **37**, 447–452, 1974b.

LAHIRI, B. N. and A. T. PRICE, Electromagnetic induction in non-uniform conductors, *Phil. Trans. Roy. Soc. London*, **A237**, 509–540, 1939.

LARSEN, J. C., Low frequency electromagnetic study of deep mantle electrical conductivity beneath the Hawaiian Islands, *Geophys. J.*, **43**, 17–46, 1975.

LILLEY, F. E. M., Analysis of daily variations recorded by arrays, *Geophys J.*, **43**, 1–16, 1975a.

LILLEY, F. E. M., Magnetometer array studies, *P.E.P.I.*, **10**, 231–240, 1975b.

LILLEY, F. E. M. and R. L. PARKER, Magnetic diurnal variation compared between eastern and western Australia, *Geophys. J.*, **44**, 719–724, 1976.

MALIN, S. C. R. and D. E. WINCH, Anomalous lunar and solar diurnal variations in the geomagnetic field, *Nature*, **218**, 941–942, 1968.

PARKER, R. L., The inverse problem of electrical conductivity in the mantle, *Geophys. J.*, **22**, 121–138, 1970.

PARKINSON, W. D., An analysis of the geomagnetic diurnal variation during the IGY, *Gerlands Beitr. Geophys.*, **80**, 199–232, 1971.

PARKINSON, W. D., The reliability of conductivity derived from diurnal variations, *J. Geomag. Geoelectr.*, **26**, 281–284, 1974.

PRICE, A. T., The induction of electric currents in non-uniform thin sheets and shells, *Quart. Jour. Mech. App. Math.*, **2**, 283–310, 1949.

PRICE, A. T., Electromagnetic induction within the earth, in *Physics of Geomagnetic Phenomena*, edited by Matsushita and Campbell, pp. 235–301, Academic Press, New York, 1967.

RICHARDS, M. L., A study of electrical conductivity in the earth, Thesis, University of California, 1970.

RIDDIHOUGH, R. P., Diurnal variation of *F* over the British Isles, *Nature*, **215**, 720–722, 1967.

RIKITAKE, T., Anomaly of geomagnetic variations in Japan, *Geophys. J.*, **2**, 176–178, 1959.

RIKITAKE, T., *Sq* and the ocean, *J. Geophys. Res.*, **67**, 2588–2591, 1962.
RODEN, R. B., The effect of an ocean on magnetic diurnal variation, *Geophys. J.*, **8**, 375–388, 1964.
SCHMUCKER, U., Anomalies of geomagnetic variations in the south-west United States, Scripps Inst. Oceanography Bulletin No. 13, 1970a.
SCHMUCKER, U., An introduction to induction anomalies, *J. Geomag. Geoelectr.*, **22**, 9–33, 1970b.
SCHMUCKER, U. and J. JANKOWSKI, Geomagnetic induction studies and the electrical state of the upper mantle, *Tectonophysics*, **13**, 233–256, 1972.
SCHUSTER, A., The diurnal variation of Terrestrial Magnetism, *Phil. Trans. Roy. Soc. London*, **A180**, 467–512, 1889.
SRIVASTAVA, B. J., D. S. B. RAO, and S. N. PRASAD, Geomagnetic induction anomalies along the Hyderabad-Kalingpatnam profile, *J. Geomag. Geoelectr.*, **25**, 247, 1974.
SUZUKI, A., A new analysis of the geomagnetic *Sq* field, *J. Geomag. Geoelectr.*, **25**, 259, 1973.

Electromagnetic Response Functions from Interrupted and Noisy Data*

J. C. Larsen

Pacific Marine Environmental Laboratory, Environmental Research Laboratories, National Oceanic and Atmospheric Administration, Seattle, Washington, U.S.A.

(Received January 30, 1978; Revised March 24, 1978)

General methods for constructing reliable electromagnetic response functions from interrupted and noisy recordings of the naturally occurring electromagnetic variations are described. The aim is to be able to effectively utilize the 11 years of hourly mean electric and magnetic data (1932–42) from Tucson, Arizona, despite the many gaps and large extraneous values. This long-time series is extremely useful in helping one to focus on techniques that yield stable and reproducible response functions. The ultimate goal is to obtain accurate response functions in order to construct reliable earth electrical conductivity models and to determine the ocean fluid-induced part of the electromagnetic field. The Tucson electric field data is found to have diurnal and semidiurnal tidal frequencies most likely of oceanographic origin and a 6 d^{-1} peak of unknown origin.

1. Introduction

This paper first describes the Tucson electromagnetic data, the probable reasons for the missing and noisy values, and the overall quality of the remaining values. Then methods are described and discussed for constructing electromagnetic response functions using the Tucson electromagnetic data as a basis for testing the validity of the methods.

2. Tucson Electromagnetic Data

In 1931, the U.S. Coast and Geodetic Survey, the Bell Telephone Company, and the Department of Terrestrial Magnetism of the Carnegie Institution of Washington began recording earth current potentials at the Tucson Magnetic Observatory, Arizona, using long (57 and 94 km) telephone cables (Rooney, 1949) with electrodes consisting of web-shaped grids of lead wire ($\sim$6 m^2 in area). During this cooperative project more than 11 years of hourly mean values were collected.

This data is unique because of its long duration and low electrode noise due to the large separation between electrodes. The utilization of this data, however, has been hampered because of noisy and interrupted data. About 4% ($\sim$160 d) of the electric field data is missing and haphazardly distributed. For example, only 37% of the data can be subdivided into 30-day segments where all components are simultaneously free of interruptions, and only 18% can be subdivided into 60-day segments. An additional 3%

* Contribution number 361 from the NOAA/ERL Pacific Marine Environmental Laboratory.

($\sim$120 d) of the data is found to have rather large errors and be randomly distributed.

Most of the missing and noisy data is reported by ROONEY (1949) to have occurred during magnetic storms and is probably the result of the electric field not being filtered prior to being recorded. That is, a magnetic storm with intense high-frequency magnetic oscillations will have even more intense electric field oscillations resulting in a possible loss of the electric field record. Even if the recording is not lost during the storm, the hourly mean values are likely to be in error due to aliasing and the difficulty of accurately hand scaling the mean values whenever the trace is rapidly oscillating. Fortunately the hourly mean magnetic values are less subject to aliasing errors and only a trivial 0.4% ($\sim$15 d) loss of data occurred. These were replaced by the mean diurnal variation based on the nearest 4 days.

The quality of the electric field data, ignoring the missing and noisy values, seems to be quite good. For example, the coherence squared between the electric and magnetic fields is 0.91 for the frequency range 0.1 to 8 d^{-1}. The spectrum of the noise in the electric field increases, however, toward the lower frequencies while the spectrum for the electric field coherent with the magnetic field decreases (Fig. 1). Thus in the frequency range 0.1 to 0.5 d^{-1}, the coherence squared falls off to 0.58. It seems to be a rough rule that the noise exceeds the signal below 0.1 d^{-1} (see Fig. 1 and LARSEN, 1975).

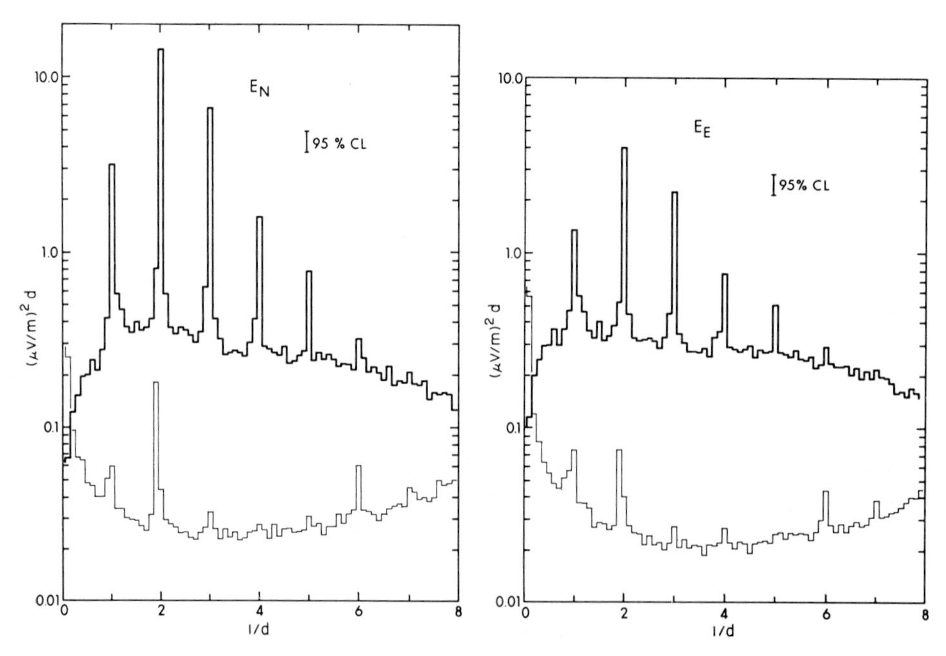

Fig. 1. Correlated (upper heavy line) and uncorrelated residual (lower thin line) spectra of north (19° T) and east (90°T) electric field components for Tucson by 0.1 d^{-1} bandwidths with 419 degrees of freedom. *T* indicates true azimuth measured clockwise from geographic north. The correlated spectrum is based on the part of the electric field coherent with the horizontal magnetic field, and the uncorrelated spectrum is from the incoherent part excluding the noisy values. Data from 11 years of hourly mean values.

The Tucson vertical magnetic data was also compared with the horizontal magnetic field for frequencies below 0.5 d^{-1}. The purpose was to calibrate the magnetotelluric response by determining the effects of surface distorting structure that may be present in the electric field but not present in the low frequency vertical magnetic field. The vertical magnetic data from Tucson is essentially complete but noisy, especially for the first few days for some intense storm time variations, D_{ST}, giving anomalous response inconsistent with the magnetotelluric E/B response. Its most likely source of noise, other than instrumental, comes from the fact that the vertical over horizontal magnetic response (V/B) is sensitive to fluctuations in the spatial extent of the source field, whereas the E/B response is not (LARSEN, 1973). Therefore it is important to remove this noise. In all, about 0.8% (~34 d) of the data was eliminated. The overall coherence squared has the low but significant level of 0.56 for the frequency range 0.1 to 0.5 d^{-1}.

3. Magnetotelluric Response (E/B)

To illustrate the method of analysis, consider a single electric $e(t)$ and magnetic $b(t)$ component for $t=-F, \ldots, F$ and let $E(f)$ and $B(f)$ be the discrete complex Fourier series. For example,

$$E(f)=F^{-1}\sum_{t=-F}^{F} e(t)\exp(i\pi ft/F) \tag{1}$$

for $f=1, \ldots, F$ frequencies.

The electromagnetic fields are related through the response function Z (m/s) by

$$E(f)=Z(f)B(f)+U(f)\,, \tag{2}$$

where U represents the uncorrelated residual noise in E and B is assumed noise-free. The situation for noise in both E and B will be discussed in a later section. The residual noise U may, in fact, contain signals caused by motionally induced electric fields due to tides, mesoscale eddies, and other long-period oscillations in the ocean. These appear as electric field noise because the magnetic fields at islands or inland generated by these motions are extremely weak compared with those generated by ionospheric sources (LARSEN, 1968). The contents of U may not, therefore, be entirely random noise at continental sites and should be examined for its oceanographic contribution. This aspect will be discussed in a later section.

4. Predicted Response Function

The E/B response can be rewritten as

$$Z(f)=X(f)Z_p(f)\,, \tag{3}$$

where Z_p is the predicted response function based on a horizontally layered conductivity model used to interpret the data at an initial stage in the analysis. The initial Z_p could, for example, be the response of an arbitrary half-space, i.e., $Z_p \sim (1-i)(f/F)^{1/2}$. The nondimensional X is the refinement of the response Z_p in order to fit the data and XZ_p can be used to construct a new Z_p. Then X will be relatively frequency-independent since Z_p is close to the final estimated Z. The predicted electric field will be $E_p(f)=Z_p(f)\,B(f)$. Thus Eq. (2) simplifies to

$$E(f)=X(f)E_p(f)+U(f)\,, \tag{4}$$

that has advantages over Eq. (2) because X will be nondimensional, nearly real, and relatively frequency-independent.

5. Band-Averaged Estimates of Response

Imagine that $X(f)$, within a given bandwidth, $f_1 \le f \le f_2$, is independent of frequency. Then the X that minimizes the following

$$\varepsilon_u^2=\frac{1}{2}\langle|E-XE_p|^2\rangle \tag{5}$$

is

$$X=\langle EE_p^*\rangle\langle|E_p|^2\rangle^{-1}\,, \tag{6}$$

where index f is implied, $\langle\ \rangle$ is the mean over $F=f_2-f_1+1$ frequencies, and E_p^* is the complex conjugate of E_p. Note that because of Z_p, X is relatively independent of frequency so that the biasing effects in estimating X from wide bandwidths are minimized.

From Eqs. (4) and (5), the variance in $|U|$ is

$$\varepsilon_u^2=\frac{1}{2}\langle|U|^2\rangle\,. \tag{7}$$

Assuming the errors in U are independent and equal for real and imaginary parts, the variance in $|X|$ will be

$$\varepsilon_x^2=F^{-1}\langle|E_p|^2\rangle^{-1}\varepsilon_u^2\,. \tag{8}$$

The variance in $|Z|$, using Eq. (3) will be

$$\varepsilon_z^2=|Z_p|^2\varepsilon_x^2\,, \tag{9}$$

and the relative error in the response $Z=XZ_p$ and X will be identical

$$r=|Z|^{-1}\varepsilon_z=|X|^{-1}\varepsilon_x\,. \tag{10}$$

For the bivariate magnetotelluric case appropriate for the Tucson magnetotelluric data, Eq. (4) becomes

$$E=X_1E_1+X_2E_2+U\,, \tag{11}$$

where E_1 and E_2 are the predicted electric components based on Z_p and the two horizontal orthogonal magnetic components. The band-averaged estimates are

$$\begin{aligned} X_1&=D^{-1}[\langle EE_1^*\rangle\langle|E_2|^2\rangle-\langle EE_2^*\rangle\langle E_1^*E_2\rangle] \\ X_2&=D^{-1}[\langle EE_2^*\rangle\langle|E_1|^2\rangle-\langle EE_1^*\rangle\langle E_1E_2^*\rangle] \end{aligned} \tag{12}$$

where

$$D=\langle|E_1|^2\rangle\langle|E_2|^2\rangle-|\langle E_1E_2^*\rangle|^2\,.$$

The variance ε_u^2 will be given by Eq. (7) with U defined by Eq. (11) and the variances in $|X_1|$ and $|X_2|$ are then

$$\varepsilon_{x_1}^2 = [(1-R_3^2)F\langle|E_1|^2\rangle]^{-1}\varepsilon_u^2$$
$$\varepsilon_{x_2}^2 = [(1-R_3^2)F\langle|E_2|^2\rangle]^{-1}\varepsilon_u^2 , \tag{13}$$

where R_3 is the coherence between E_1 and E_2. The variance and relative error in the responses are then computed, respectively, by Eqs. (9) and (10).

Assuming the horizontal magnetic data is noise free, the confidence interval for Z, using the Student's t distribution (JENKINS and WATTS, 1969), is

$$L_z = \alpha\varepsilon_z , \tag{14}$$

where $\alpha \approx 2$ for the 95% confidence interval for $2F > 20$ degrees of freedom. The standard error is ε_z and is approximately equal to the 68% confidence interval.

6. Smooth Estimates of Response

6.1 Power series representation

When the predicting models are close to the final model that we seek, X will be nearly real and frequency-independent. Therefore X can be represented by a smoothly varying function of frequency independent of wavenumber because the ionospheric wavelengths are normally large compared with the penetration depth (LARSEN, 1973). We therefore assume a power series representation,

$$X(f) = \sum_{n=0}^{N-1} A_n(f/F)^{n/2} , \tag{15}$$

for $f=1, \ldots, F$. The advantages of an expansion of this sort rather than band-averaged estimates are the following: (1) The response will be a smoothly varying function of frequency rather than a staircase-type response implied by band-averaged estimates. This eliminates any subsequent smoothing often demanded when using band-averaged estimates. (2) The decision about the size of the bandwidths is eliminated because the response is determined for the entire frequency range. (3) The response can be readily interpolated at certain frequencies, such as the tides, where estimates of the response may be strongly distorted by spectral lines in the noise. This interpolation feature is important for determining the ocean fluid-induced contribution. (4) The smoothing improves the ease with which the response can be used to construct conductivity models. (5) Once the coefficients of the power series representation are determined, the response can be digitized for any frequency interval. This simplifies conductivity model construction. (6) Smoothing makes it possible to easily compute derivatives with respect to frequency of the response function. These are useful to test whether the response can be interpreted by a horizontally layered conductivity model (WEIDELT, 1972).

Care must be taken, however, that the power series representation remains stable by not allowing N, the number of coefficients, to become too large so that false oscillations are introduced. On the other hand, N must be large enough to avoid any severe distortion due to smoothing. These difficulties are minimized by a careful choice of (1) the predicted response Z_p, (2) the functional form of X, and (3) weights. Then N need not be large. We have already discussed the selection of Z_p, and the weights will be discussed later. The preferred form of X was found by comparing a power series expansion in frequency, square root frequency, and Fourier cosine series for a given $N=10$ and selecting the functional form that minimizes the sum

$$\sum_{f=1}^{F} |X(f)-Z_1(f)/Z_2(f)|^2 ,$$

where Z_1 and Z_2 were two different response functions used to approximate the Tucson E/B response for the second and third cycle in the data analysis and interpretation. For this case the power series expansion in square root frequency was clearly the superior representation when the number of coefficients was limited to N less than 10. This particular form also seems appropriate because the square root of the frequency occurs naturally in the electromagnetic response functions. For example, the response of a uniformly conducting half-space is $Z=(1-i)(\omega/2\mu\sigma)^{1/2}$, where ω is the radian frequency, σ the conductivity, and μ the magnetic permeability.

It might justifiably be argued that any power series expansion will be inadequate because of the inappropriateness of a power series representation and because of the unknown effects of smoothing implied by the use of a finite N. What the best form of X should be is not known, but clearly it must be at least general enough to include possibilities of conductivity models that are not necessarily horizontally layered.

Some indication of the smoothing can be gotten by comparing the power series representation of the Tucson response functions at different frequencies for various values of N. Table 1 illustrates the effects of N on the effective response $\bar{Z}=(Z_{yx}-Z_{xy})/2$. The Z_{xy} and Z_{yx} are the off-diagonal terms of the two-by-two response matrix for the bivariate case that consists of comparing each horizontal electric component with the two simultaneous horizontal magnetic components. The results in Table 1 show that the power series representation has converged, for $N>6$, to a common response independent of N. Note the excellent agreement to the mean based on $N=7$ to 10 (Table 1). For smaller values of N the results show a distortion due to smoothing. At the highest fre-

Table 1. Effective Tucson magnetotelluric response $\bar{Z}=(Z_{yx}-Z_{xy})/2$ in meters per second as function of frequency and N, the number of terms in the power series representation. Mean is based on $N=7$ to 10. Band estimates from 1 d^{-1} bandwidth. Z_p is predicted response.

N	$\bar{Z}$ (0.5 d^{-1})		$\bar{Z}$ (1.5 d^{-1})		$\bar{Z}$ (4.5 d^{-1})		$\bar{Z}$ (7.5 d^{-1})	
	Real	−Imag	Real	−Imag	Real	−Imag	Real	−Imag
1	22.6	45.9	54.2	81.8	77.8	167.3	92.7	260.9
2	13.7	51.2	46.6	87.1	91.5	156.7	146.6	230.8
3	12.8	41.4	46.2	86.3	90.9	159.9	146.8	181.1
4	15.2	39.0	44.2	86.7	95.4	158.0	131.0	187.6
5	15.5	39.6	43.9	86.3	95.2	156.6	131.1	193.2
6	15.6	38.7	43.9	86.9	95.3	156.1	131.9	190.3
7	15.6	39.1	43.8	86.8	95.3	155.0	131.9	189.5
8	15.7	39.3	43.8	86.7	95.2	154.7	131.9	189.6
9	15.7	39.2	43.8	86.7	95.1	154.9	132.0	189.4
10	15.6	39.3	43.8	86.7	95.0	155.0	132.2	189.0
Mean	15.6	39.2	43.8	86.7	95.2	154.9	132.0	189.4
Band	16.2	42.5	45.9	88.2	95.0	153.6	132.6	189.7
Z_p	19.8	38.8	47.1	68.9	68.3	141.5	82.2	220.9

quency (7.5 d^{-1}) the response does not converge as rapidly as it does at the lower frequency. The probable cause is that Z_p (Table 1) is not as close to the final $\bar{Z}$ at the higher frequencies. The power series representation also compares favorably to the band-averaged estimates for 1 d^{-1} bandwidths (Table 1). This provides an additional check on the stability of the power series representation. In fact the confidence intervals on the band-averaged estimates can be conveniently used as the confidence intervals on the power series representation.

The above results show that it is possible to choose a Z_p such that the power series expansion of X is stable, relatively distortion-free, and rapidly and smoothly convergent at each frequency.

6.2 *Response in time domain*

In the time domain Eq. (4) implies the convolution

$$e(t)=(2F)^{-1}\sum_{\tau=-F}^{F} e_p(\tau)x(t-\tau)+u(t)\,. \tag{16}$$

The use of Fourier transforms of finite length time series implies that $x(t)=x(t+2F)=x(t-2F)$ and $e(t)=e(t+2F)=e(t-2F)$. Thus if the effective width of $x(t)$ is broad, there will be end values, for example, $e_p(-F+m)$ for $0<m\ll F$, that will be used in Eq. (16) to compute $e(F)$ when in fact what should be used, if the series were longer, is $e_p(F+m)$. The importance of a carefully chosen Z_p is now apparent because $X(f)$ will then be nearly real and frequency-independent which implies that $x(t)$ will have a narrow symmetric sharp peak. Because there will always be some width to $x(t)$, the end points of the time series are de-emphasized by applying a cosine taper $0.5-0.5\cos[\pi(|t|-F)/M]$ for $|t|=F-M, \ldots, F$ where $M=10$ is found to be of convenient length. The point here is that M should be greater than the effective width of $x(t)$.

6.3 *Power series coefficients*

The A_n's in Eq. (15) are determined by finding the minimum of

$$\varepsilon_u^2=\frac{1}{2}\langle W(f)|E(f)-\sum_{n=0}^{N-1} A_n(f/F)^{n/2}E_p(f)|^2\rangle\,, \tag{17}$$

where the choice of the weights W are discussed in the next section. Initially we assume the weights have equal value. The complex A_n's are found from the set of linear equations

$$\sum_{n=0}^{N-1} A_n\langle W|E_p|^2(f/F)^{(n+m)/2}\rangle=\langle WEE_p^*\,(f/F)^{m/2}\rangle\,, \tag{18}$$

for $m=0, \ldots, N-1$ where the solution is found by matrix inversion. In order to improve the stability of the inversion by the method of damped least squares, the diagonal terms, $\langle W|E_p|^2(f/F)^n\rangle$, of the matrix are increased by a small factor, 10^{-8}. The instability arises from the very small eigenvalues that are likely to occur if there is noise. Increasing the diagonal matrix terms by a very small amount increases the value of the eigenvalues so that the inverse, which involves the reciprocal of the eigenvalues, is not overwhelmed by enormously large but extraneous values. The generalization of Eq. (18) to the bivariate case, see Eq. (11), is readily apparent and will not be described here.

6.4 *Weights*

Expression (17) can also be written as

$$\varepsilon_u^2 = \frac{1}{2} \langle W(f) |U(f)|^2 \rangle \,. \tag{19}$$

The method of least squares requires the residual noise U to be white. In other words, the weights should be chosen such that all portions of E are equally emphasized with the further requirement that the weights be smoothly varying in frequency so that only a few degrees of freedom are utilized. Essentially the weights whiten the residual spectrum by removing the trend and eliminating any spectral lines so that the entire frequency range is effectively utilized in computing the response.

The effects of the spectral lines can be eliminated by the following method. Each residue $R(f) = |U(f)|^2$ from a band of 30 frequencies is compared to the variance $\varepsilon_u^2 = \langle |U|^2 \rangle$ for that band. The weight $W(f)$ is then set to zero whenever $R(f) > 5.9\,\varepsilon_u^2$ because the residue is presumably periodic. There is a 5% chance, out of a sample of 30 random residues, that a residue will exceed this level and therefore be misidentified as periodic (FISHER, 1929).

The trend in the residual spectrum is then estimated by fitting a polynomial, $P(f) = \sum_{n=0}^{N} B_n (f/F)^n$, to $2 \log |U(f)|$ excluding the spectral lines. The weights will be $W(f) \sim \exp(-P(f))$. The value of N is determined by examining whether the weighted residual spectrum $\langle W|U|^2 \rangle$ is a reasonable approximation to white noise. For Tucson, $N=3$ is found to be sufficient. Finally the mean of the weights are normalized to unity, $\langle W \rangle = 1$.

7. Electric Field Data Gaps

Let the gap function, $g(t)$ for $t = -F, \ldots, F$, be unity for the data and zero for the missing data. The available time series is

$$e_g(t) = g(t) e(t) \tag{20}$$

rather than $e(t)$. In the frequency domain,

$$E_g(f) = G(f) * E(f) \tag{21}$$

for $f = 1, \ldots, F$, where G is the Fourier transform of $g(t)$ and $*$ represents the convolution integral. Then instead of Eq. (4) the response is defined by

$$E_g(f) = G(f) * [X(f) E_p(f)] + U_g(f) \,, \tag{22}$$

where the knowns are E_g, E_p, and G. The unknown X is the value that minimizes

$$\langle W(f) | E_g(f) - G(f) * [X(f) E_p(f)] |^2 \rangle \,, \tag{23}$$

where X is represented by the power series Eq. (15). To solve Eq. (23) efficiently an iterative scheme is used. First, X is assumed to be nearly frequency-independent because a good choice for Z_p has been made. Thus X can be brought outside the convolution integral so that the first estimate of X is defined by

$$E_g \approx (G * E_p) X^1 + U_g^1 \tag{24}$$

and found by least squares as described in Section 6. Then the estimates of X are refined by the following iterative scheme:

$$E_g \approx [G*(X^m E_p)](X^{m+1}/X^m) + U_g^{m+1} \tag{25}$$

for $m=1, \ldots, M$, iterations where X^{m+1} is the $m+1^{th}$ least squares solution using the mth solution X^m. The iterations are stopped when the coherence squared,

$$R_m^2 = \langle |E_g - U_g^m|^2 \rangle \langle |E_g|^2 \rangle^{-1}, \tag{26}$$

ceases to vary by 1%. For the Tucson data this usually occurs within a few iterations.

The convolution integral is constructed by two applications of the fast Fourier transform; one to convert $X^m(f)E_p(f)$ to $e^m(t)$, and a second to convert $g(t)e^m(t)$ to $G(f)*E^m(f)$.

8. Gross Electric Field Errors

The power series representation of X permits us to compute $W(f)^{1/2}U_g(f)$ and the weighted time series $u_w(t)$. The gross errors are identified as those values in $u_w(t)$ that exceed four times the rms value of the series. In the Tucson data large errors seem to occur during intense magnetic storms when the electric field recordings are often lost. As already discussed, these errors probably would not have occurred if the data had been filtered prior to being recorded. In order to eliminate the effects of these errors, $g(t)$ is set to zero for these values and for the intervening value if two noisy data points are separated by a single data point. In the iterative scheme for determining $X(f)$, the noisy values are examined at each iteration and the values of $g(t)$ and $G(f)$ changed. The convergence, however, is still found to be rapid and smooth with the number of iterations usually being less than ten. The weights are also rocomputed for each cycle.

The Tucson electric field data is found to have only about 3% extremely noisy data. The effects of removing these values, however, are quite pronounced. For example, in some portions of the data it is impossible to detect any tidal signals in the residuals prior to removing the noisy values. Some of the rejected data may be due to noise in the magnetic field. However, it probably does not matter much whether the rejected noise is misidentified as being in the electric field for the following reasons. Suppose the noise in the magnetic field consists of a single large spike. Then the electric field time series will, by Eq. (16), contain noise smeared out over several data points. However, this noise, if it is large, will be rejected. Thus the principal part of the noise is eliminated and the main effects of misidentifying it as electric field noise are to cause extra data to be rejected and to cause some minor noise to remain because of the smearing effect.

9. Noisy Electric and Magnetic Data

Up to now it has been assumed that all the noise lies in the electric field. This assumption is essentially valid because the electric field data was not prefiltered and because a near coastal site is much more sensitive to ocean-induced electric fields (here treated as noise) than to ocean-induced horizontal magnetic fields. This does not preclude, however, the possibility of some noisy magnetic data which is discussed in the following.

For noise-free magnetic data Eq. (4) is used with band-averaged response given by Eq.(6). For noise-free electric data we have instead

$$E_p(f)=YE(f)+U(f) \tag{27}$$

with band-averaged response

$$Y=\langle E^*E_p\rangle\langle|E|^2\rangle^{-1}. \tag{28}$$

Now let the noise be random so that the noise in E and E_p are uncorrelated. Then the cross-spectrum $\langle EE_p^*\rangle$ is not biased by the noise but the spectra will be by positive real factors $\alpha\leq 1$ and $\beta\leq 1$ with unbiased spectra $\alpha\langle|E|^2\rangle$ and $\beta\langle|E_p|^2\rangle$. The unbiased estimates of X and Y are then $\bar{X}=\alpha^{-1}X$ and $\bar{Y}=\beta^{-1}Y$. Since $\bar{X}\bar{Y}=1$, we have $\alpha\beta=\langle EE_p^*\rangle\langle E^*E_p\rangle\langle|E|^2\rangle^{-1}\langle|E_p|^2\rangle^{-1}$. The right-hand side is merely the coherence square between E and E_p; that is, $\alpha\beta=R^2$, and α and β are restricted to $R^2\leq\alpha\leq 1$ and $R^2\leq\beta\leq 1$. The estimate of the amplitude of the response lies, therefore, within the limits

$$|X_{\min}|\leq|\bar{X}|\leq|X_{\max}| \tag{29}$$

where $X_{\min}=\langle EE_p^*\rangle\langle|E_p|^2\rangle^{-1}$ and $X_{\max}=\langle|E|^2\rangle\langle EE_p^*\rangle^{-1}$. The phase is uniquely determined. The above discussion shows that $X_{\max}=X_{\min}/R^2$ or

$$X_{\max}=X_{\min}(1+\langle|U|^2\rangle\langle|C|^2\rangle^{-1}) \tag{30}$$

where U is the uncorrelated residual noise in E and C is the correlated part of E. One sees, therefore, that the spread in X clearly depends on the ratio of signal to noise.

These limits will not define X very closely if the coherence is low. Therefore the sensitivity of α and β to the noise needs to be determined. Assume therefore that $Z_p\approx Z$ so that $\langle|E|^2\rangle\approx\langle|E_p|^2\rangle$. Then the variance in E divided by the variance in E_p is $r=(1-\alpha)/(1-\beta)$. The solution for α, given r and $\beta=R^2/\alpha$, is then

$$\alpha=(1-r)/2+[(1-r)^2/4+rR^2]^{1/2}. \tag{31}$$

For $r>9$ we find $\alpha\approx R^2$ and $\beta\approx 1$; i.e., E_p is essentially noise-free, and when $r<1/9$ we find $\alpha\approx 1$ and $\beta\approx R^2$ with E essentially noise-free. For example, if $R^2=0.5$ and the rms noise in E is three times that in E_p, then $r=9$ and $\alpha=0.53\approx R^2$ and $\beta=0.95\approx 1.0$. Thus E_p can be treated as almost noise-free. Therefore if it can be assumed that the rms noise is at least thrice as large in one series as in the other, then one series can be treated as essentially noise-free. If, on the other hand, we have no knowledge about the quality of either the electric or magnetic data but suspect that both series are equally noisy, then equal relative error can be assumed, i.e., $\alpha=\beta=R$, $\bar{X}=(X/Y)^{1/2}$, $X_{\min}=\bar{X}R$, and $X_{\max}=\bar{X}/R$.

In the present paper the horizontal magnetic field components are, quite properly, assumed to be almost noise-free except for a small random white noise, $\varepsilon_b\approx 1$ nT. The correction of the magnetic spectrum for this noise is then determined from the following relationship based on Parseval's theorem:

$$\varepsilon_B^2=(1-\beta)F\langle|B|^2\rangle. \tag{32}$$

where F is the number of frequencies within the band and ε_B the rms noise. The correction β is found, however, to differ from unity by less than 0.1%.

For the bivariate magnetotelluric case utilizing two horizontal magnetic components, Eq. (11), the unbiased estimates $\bar{X}_1$ and $\bar{X}_2$ are derived from Eq. (12) using unbiased spectra $\alpha\langle|E|^2\rangle$, $\alpha_1\langle|E_1|^2\rangle$, and $\alpha_2\langle|E_2|^2\rangle$ where the α's are all less than unity. Starting with $E=\bar{X}_1E_1+\bar{X}_2E_2$, we find, after some manipulation, that

$$\frac{|R_1|^2}{\alpha\alpha_1}+\frac{|R_2|^2}{\alpha\alpha_2}+\frac{|R_3|^2}{\alpha_1\alpha_2}-\frac{2\ \mathrm{Real}\ (R_1R_2^*R_3)}{\alpha\alpha_1\alpha_2}=1 \tag{33}$$

where the complex coherences are

$$R_1=\langle EE_1^*\rangle[\langle|E|^2\rangle\langle|E_1|^2\rangle]^{-1/2}$$

$$R_2=\langle EE_2^*\rangle[\langle|E|^2\rangle\langle|E_2|^2\rangle]^{-1/2}$$

$$R_3=\langle E_1E_2^*\rangle[\langle|E_1|^2\rangle\langle|E_2|^2\rangle]^{-1/2}\ .$$

For equal relative noise, $\alpha=\alpha_1=\alpha_2$, the solution is found from the cubic equation

$$\alpha^3-(|R_1|^2+|R_2|^2+|R_3|^2)\alpha+2\ \mathrm{Real}\ (R_1R_2^*R_3)=0\ . \tag{34}$$

10. Residual and Oceanic Motional Induced Electric Field

The uncorrelated residual electric field spectrum at Tucson (Fig. 1) reveals prominent peaks at 0.9, 1.9, and 6.0 d^{-1} for 0.1 d^{-1} bandwidths. The spectra have been approximately corrected for the effects of the gaps and rejected values by dividing the spectra by $G^2(0)=\langle g\rangle^2$. This is the proper correction when the gaps are randomly distributed. A higher resolution spectrum with 0.025 d^{-1} bandwidths (Fig. 2) shows that the peak at 0.9 d^{-1} seems to consist principally of the tidal constituents O_1 (0.93 d^{-1}) and K_1 (1.00 d^{-1}), and the peak at 1.9 d^{-1} consists principally of M_2 (1.93 d^{-1}). A comparison of these peaks by seasons shows no seasonal variation. These spectral peaks are, therefore, most likely due to tidally induced electric currents originating in the Gulf of California (200 km distance) and in the Pacific Ocean (450 km distance). The upper Gulf of California seems the preferred source because the tidal constituent ratio $S_2/M_2\sim0.6$ observed there is more in agreement with the electric field than is the Pacific tidal constituent ratio $S_2/M_2\sim1.0$. The stability of these tidal peaks gives us some assurance that the method of analysis is valid. The origin of the 6.0 d^{-1} peak is unknown.

The background residual noise is about 10% of the energy of the correlated signal for most of the frequency range. Only near and below 0.1 d^{-1} does the residual noise begin to dominate. This rise in the residual spectrum toward the low frequencies may be due to instrumental and recording noise, but the presence of the prominent tidal peaks suggests the possibility that some of the low-frequency electric field may be induced by ocean oscillations. The presence of these signals may even be the limiting factor for accurate deep electromagnetic soundings at continental sites such as Tucson. On the other hand, it may be possible to use data from sites such as Tucson to study low-frequency large-scale oscillations in the ocean.

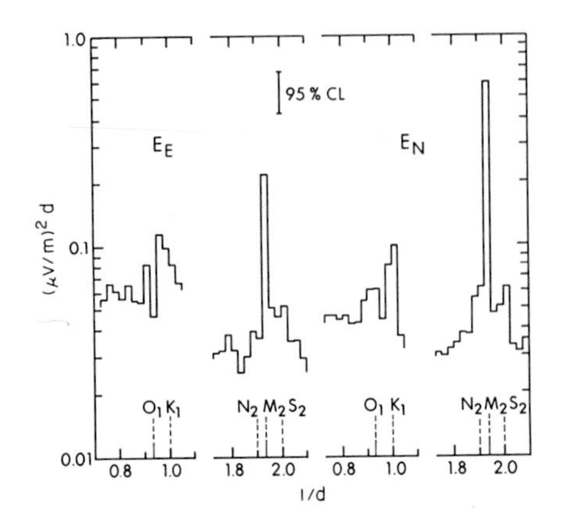

Fig. 2. High resolution uncorrelated residual spectra of east (90° T) and north (19°T) electric field components by 0.02 d^{-1} bandwidths with 168 degrees of freedom for the diurnal and semidiurnal tidal frequency range.

11. Magnetic Variational Response (V/B) and Surface Conductivity Effects

The V/B response for low frequencies $<0.5\ d^{-1}$ has the advantage of not being distorted by surface conductivity effects that are usually present in the electric field data. If both V/B and E/B responses are available over a range of frequencies $<0.5\ d^{-1}$, it will be possible to determine and correct for the surface effect in E/B. Here we describe how the V/B response can be used to determine the surface effect in the generally more accurate E/B response with its usually wider useful frequency range.

With a simple electromagnetic wave impinging on a flat surface, $V=(k/\omega)E$ (LARSEN, 1973) where E is the horizontal electric field normal to the propagating direction, V is the vertical magnetic field, and k is the horizontal wavenumber. Then Z', the V/B response, will be related to Z, the E/B response, by

$$Z'=(k/\omega)Z\,. \tag{35}$$

The E/B response is insensitive to the wavenumber provided k is small, i.e., $ka<6$ (a=Earth's radius) (LARSEN, 1973), but Eq. (35) shows that the V/B response is quite sensitive to wavenumber and its fluctuations. The coherence between V and B for the background continuum is observed to be usually much less than that between E and B for sites such as Tucson where the vertical magnetic field is a small fraction of the horizontal magnetic field.

In order to minimize the effects of fluctuations in the wavenumber, one selects and analyzes only special magnetic events. Here we consider the storm time, D_{ST}, variations that seem to describe the magnetic fluctuations for frequencies between 0.05 and 0.5 d^{-1} (BANKS, 1969). The D_{ST} variations are describable (BANKS, 1969) by a geomagnetic zonal ionospheric electric current with geomagnetic latitudinal variation, $\sin\theta$, and a single magnetic north component, B. The V/B response is related to the E/B response (SCHMUCKER, 1970, p. 59) by

$$Z'(f)=(2i\cot\theta/\omega a)Z(f)\,, \tag{36}$$

where ω is the radian frequency, a is the Earth's radius, and θ=40.4°N is the geomagnetic latitude for Tucson. The predicted vertical magnetic field will be

$$V_p(f)=(2i \cot \theta/\omega a)Z(f)B(f)\,, \tag{37}$$

where Z is the E/B response determined from the magnetotelluric data containing a possible surface effect. The E/B response corrected for surface effects will be $DZ(f)$ where we seek a D that is real and frequency-independent (LARSEN, 1975). The vertical and predicted vertical magnetic components including noise will be related by

$$V(f)=DV_p(f)+U(f). \tag{38}$$

The value of D is then computed from

$$D=\langle VV_p^*\rangle\langle|V_p|^2\rangle^{-1}\,, \tag{39}$$

where the noise is assumed to be in the vertical component and the frequencies are from 0.1 to 0.5 d^{-1}.

For the Tucson data we use the effective magnetotelluric response $\bar{Z}$ as described in Section 6 and partially listed in Table 1. It is found, however, that there are portions of the Tucson data for which the estimate of D is drastically different from the mean with a large imaginary component that makes the response DZ completely inconsistent with any model used to interpret Z. These anomalous values of D come from very energetic but small portions of the data. These portions are identified and eliminated as in Section 8 whereby any value exceeding four times the rms value of the weighted residual time series is eliminated. When 0.8% (~34 d) of the data is eliminated from 11 years of data, the value of D is found to be comparable between year segments (Table 2) and to have the overall value real $(D)=0.37$ and imag $(D)=0.00$. Hence D is real as advertised. If all the noise is assumed to be in B, then real $(D)=0.68$ and imag $(D)=0.01$.

The portions of the data that give rise to the anomalous D are found to occur during the first few days of the initial stage of some of the intense D_{ST} variations. For example, Fig. 3 shows that the first four days of a particularly large D_{ST} variation give large residues. This suggests that the initial stage of D_{ST} variations for some individual storms are not

Table 2. Surface conductivity parameter D by yearly segments and 11-year total assuming noise-free horizontal magnetic field D (min) and noise-free vertical magnetic field D (max) for frequency band 0.1 to 0.5 d^{-1}.

Year	D (min)		D (max)		% Rejected series
	Real	Imag	Real	Imag	
1932	.35	.00	.89	−.01	0.9
1933	.35	.08	.90	.22	1.3
1934	.39	−.01	1.18	−.04	0.8
1935	.39	−.02	.87	−.05	0.1
1936	.39	.04	.77	.08	0.0
1937	.36	−.04	.67	−.07	1.9
1938	.39	.03	.66	.05	0.0
1939	.35	−.03	.53	−.04	0.3
1940	.41	−.01	.74	−.02	0.4
1941	.36	.06	.61	.11	2.9
1942	.30	.01	.67	.02	0.3
1932–42	.37	.00	.68	.01	0.8

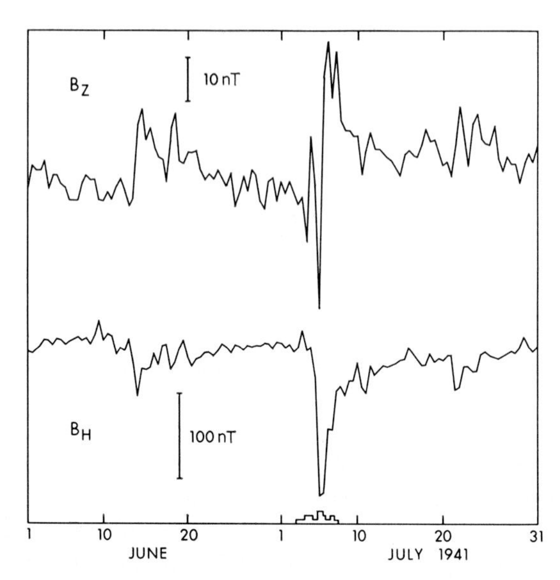

Fig. 3. Two-month low-passed (<0.85 d^{-1}) example of vertical (down) magnetic component (upper curve) and north (6° T) magnetic component (middle curve) during an intense, D_{ST}, storm time variation. Rejected values (bottom staircase curve) represented by 10 nT increments of absolute value of residuals. Maximum rejected residual was 23 nT, and nonrejected residuals were less than 3 nT.

describable by a sin θ latitudinal variation. That is, the wavenumber of the source field varies in time. This clearly indicates the need to carefully examine the individual D_{ST} variation before computing the V/B response.

12. Conclusions

The important points to consider in computing electromagnetic response functions are the following: (1) Weights are required to prewhiten the residual noise. (2) Predicted response functions Z_p are useful because the correction function X will then be nondimensional, nearly real, and relatively independent of frequency. (3) Smooth response functions are possible by power series expansions of X. (4) Short wavelength ionospheric source fields can occur and it is important to identify and eliminate their effects. (5) The effects of gaps and very noisy values can and must be removed from the data. (6) The remaining noise can be distributed between electric and magnetic components but only in an arbitrary way without additional information. (7) Oceanic motional induced fields are detectable at continental sites such as Tucson and must, therefore, be considered.

REFERENCES

BANKS, R. J., Geomagnetic variations and the electrical conductivity of the upper mantle, *Geophys. J. R. Astr. Soc.*, **17**, 457–487, 1969.

FISHER, R. A., Tests of significance in harmonic analysis, *Proc. Roy. Soc.* **A125**, 54–59, 1929.

JENKINS, G. M. and D. G. WATTS, *Spectral Analysis and Its Application*, 525 pp., Holden-Day, San Francisco, 1969.

LARSEN, J. C., Electric and magnetic fields induced by deep sea tides, *Geophys. J. R. Astr. Soc.*, **16**, 47–70, 1968.

LARSEN, J. C., An introduction to electromagnetic induction in the ocean, *Phys. Earth Planet. Inter.*, **7**, 389–398, 1973.

LARSEN, J. C., Low frequency (0.1–6.0 cpd) electromagnetic study of deep mantle electrical conductivity beneath the Hawaiian Islands, *Geophys. J. R. Astr. Soc.*, **43**, 17–46, 1975.

ROONEY, W. J., Earth-current results at Tucson magnetic observatory 1932–1942, *Carnegie Inst. Washington, Dept. Terr. Magn. Res.*, **9**, 1–309, 1949.

SCHMUCKER, U., Anomalies of geomagnetic variations in the Southwestern United States, *Bull. Scripps Inst. Oceanogr.*, **13**, 1–165, 1970.

WEIDELT, P., The inverse problem of geomagnetic induction, *Z. Geophys.*, **38**, 257–289, 1972.

Deep Conductivity Distribution on the Russian Platform from the Results of Combined Magnetotelluric and Global Magnetovariational Data Interpretation

A. A. KOVTUN and L. N. POROKHOVA

Physical Institute, University of Leningrad,
Leningrad, U.S.S.R.

(Received November 20, 1977; Revised April 4, 1978)

Local magnetotelluric sounding data from the Russian Platform are interpreted jointly with global magnetic data of long periodic variations. In the first stage of interpretation the magnetotelluric data are used to find for each site a layered conductivity model for the sediments and the underlying crust and uppermost mantle, regarding the deeper upper mantle as a uniform conductor. In the second stage conductivity estimates are derived down to 1,200 km depth from global data, using for the top layers the results from stage one.

The interpretation follows the statistical approach of the NEWTON-LeCAME-MARQUARDT method which takes a random component in the data into account. The best fitting model involves a conductivity increase by two orders of magnitude in 340 km depth, even though this result should not be regarded as final and unambiguous.

It has been established from study of magnetotelluric sounding (MTS) curves obtained in the north-west of the Russian Platform (KOVTUN, 1976) that there are a number of regions of considerable spatial extent, where the MTS curves show a good agreement with the global magnetotelluric curve derived from geomagnetic variations data according to BERDICHEVSKY *et al.* (1969). There exist at least three regions where such a close agreement of the MTS curves with the global magnetotelluric curve is observed. These are the eastern and the southern slopes of the Middle Russian Depression, the central part of the Latvian Saddle including its sides, and the northern slope of the Baltic Depression. Figure 1 shows the sounding longitudinal curves which are least distorted by the "S-effect", obtained at the northern side of the Baltic Depression (BERDICHEVSKY *et al.*, 1973). In the same figure the values of ρ_T are given, derived from the measurements of magnetic variations. These values of ρ_T were calculated by ROKITYANSKY (1975), BORISOVA *et al.* (1974), JODENTSCHYKOVA (1975) from the results of spherical harmonic analysis of the variations of *Sq*, *Dst*, and *N* types. The coincidence of curves obtained from the magnetotelluric sounding and the geomagnetic variations may indicate that the geoelectric conditions throughout the investigated regions are close, beneath a certain level, to average conditions for the whole continent. This assumption makes it possible to carry out an interpretation of the magnetotelluric sounding curves together with the global magnetotelluric curve. Earlier a similar approach was applied SCHMUCKER and YANKOWSKY

(1972), VANJAN *et al.* (1977). The broadening of the range of periods increases the information content of the data and enables us to make more definite suggestions about the distribution of conductivity in the Earth's crust and mantle. The latter possibility seems to be of especial interest. As a study and interpretation of the informative intervals show (KOVTUN and POROKHOVA, 1974), the geomagnetic variation data alone permit the Earth's electrical conductivity distribution to be estimated at levels deeper than 200 km.

With respect to better separation of the resistivities of the upper crust and mantle, a region having small longitudinal conductivity of the sedimentary cover is the most interesting. On this account the main attention in this paper will be paid to the joint interpretation of curves obtained in the western part of the Russian Platform, in the region of Latvian Saddle and on the northern slope of Baltic Depression. The problem will be solved within the framework of the spherical layered model Earth with the assumption that the conductivity distribution is a piecewise continuous function $\sigma_1, \sigma_2, \ldots, \sigma_N$, where N is the number of layers.

In the general case we have to find $\sigma_1, \sigma_2, \ldots, \sigma_N$ and thickness of corresponding layers from the MTS data set obtained over a relatively large area (300×300 km^2) and the global MTS curve plotted from the geomagnetic variations data, provided that the solution of the direct problem is known for the spherical harmonic of number $n=1$ (SOCHELNIKOV, 1968):

$$\rho_T(\vec{\theta}, T) = \frac{1}{\omega\mu_1} |Z_N(\vec{\theta}, T)|^2 \tag{1}$$

$$Z_N(\vec{\theta}, T) = \frac{-i\omega\mu_1}{k_1} \cdot \left[\frac{1}{R_N - 1/k_1 z_1} - \frac{1}{k_1 z_1} \right]^{-1} \tag{2}$$

where

$$R_{N-n+1} = \mathrm{cth}\left[k_n h_n + \mathrm{arcth}\left(\frac{1 + k_n^2/k_{n+1}^2}{k_n z_{n+1}} + \frac{k_n}{k_{n+1}} R_{N-n} \right) \right],$$

$$n = 1, 2, \ldots, N,$$

$$R_1 = \mathrm{cth}(k_n z_n),$$

$$k_n = \sqrt{i\mu_n \omega / \rho_n} \text{ and } \omega = 2\pi/T.$$

T is the period of the field variation, μ_n the magnetic permeability of the medium, h_n, z_n, and $\rho_n = 1/\sigma_n$ are the thickness, the radius, and the resistivity of the n-th layer, respectively. $h_n = z_n - z_{n+1}$, z_1 is the Earth's radius, and $\vec{\theta}$ the vector of the profile parameters σ_n, h_n. In the limit of small wave-length ($|k_1 z_1| \gg 1$) formula (1) reduces to the well-known formula of CAGNIARD (1953) which is valid for plane Earth surface and uniform wave. The problem is solved using known algorithms for the interpretation of the MTS curves based on methods of mathematical statistics (POROKHOVA, 1971, 1975). This means that the existence of a random component in the data ρ_T, which are to be interpreted, is taken into account.

The algorithm of the joint interpretation was tested using a set of curves obtained at the northern slope of the Baltic Depression, where sedimentary cover has small longitudinal conductivity. Data from several soundings are shown in Fig. 1.

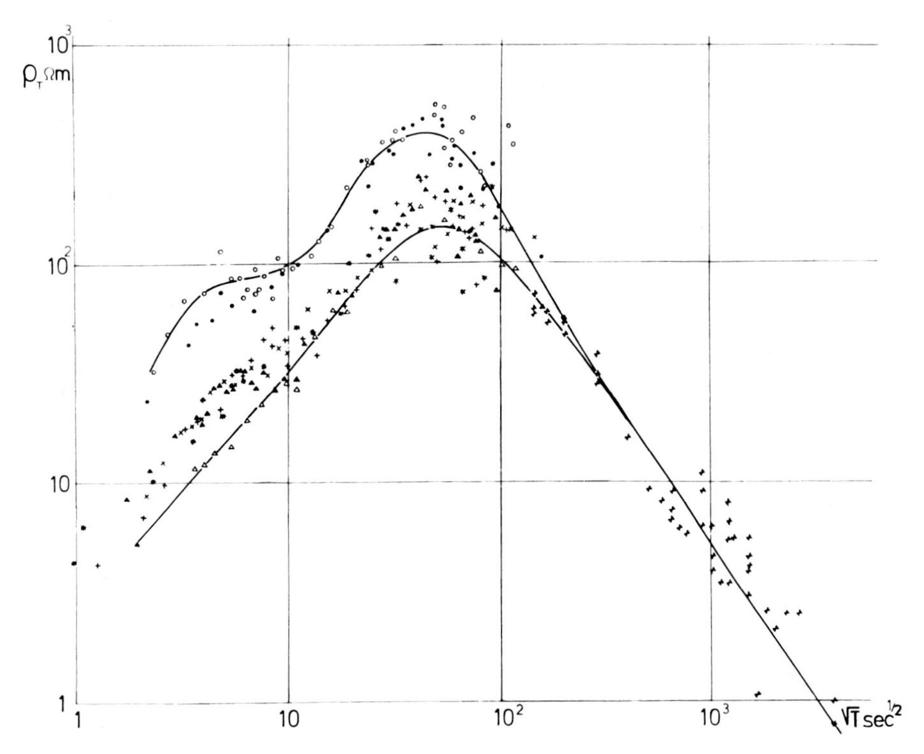

Fig. 1. MTS data at several points of the northern slope of Baltic Depression and the global sounding curve from the magnetic variations data. Legend for plots: MTS data at Dundaga—×, at Kandava—○, at Spara—▲, at Gaiki—●, at Adse—*, at Vendsava—△, at Rinda—+, global MTS data—✦. Solid lines—results of computation of MTS curves for Vendsava and Kandava in case of model 1.

The solution was obtained in two stages. In the first stage the interpretation of data sets had been performed separately for each point, which made possible the evaluation of the parameters of the sedimentary formation and the upper surface of the mantle. As an initial approximation, the following values of the profile parameters were assumed (KOVTUN *et al.*, 1965): $h_1=1.63$ km, $\rho_1=5$ Ohm m; $h_2=30$ km, $\rho_2=1{,}000$ Ohm m; $h_3=70$ km, $\rho_3=200$ Ohm m; $h_4=300$ km, $\rho_4=5{,}000$ Ohm m, $\rho_5=50$ Ohm m.

Based on the results of the examination of the information content of the MTS curves obtained in the investigated region (KOVTUN and PROKOHOVA, 1974), we have fixed the values of the parameters ρ_1, ρ_2, and ρ_4, as the interpreted field does not contain any information on the resistivity of the sedimentary cover and, in fact, does not depend on the value of the resistivity of the poorly conducting basement. The number of unknown parameters was thus decreased. This made it possible to obtain more credible values for them. The results of the interpretation are summarized in Table 1.

In the second stage data from all the sounding points together with published data on the global MTS curve (ROKITYANSKY, 1975; BORISOVA *et al.*, 1974; JODENTSCHYKOVA, 1975) were taken as information set to be interpreted.

Table 1. Results of the separate interpretation of the magnetotelluric data. Thickness in km, resistivity in Ohm m.

Points	Thickness in km				Resistivity in Ohm m	
	1	2	3	4	3	5
Priekuli	1.0	0.45	89	276	182	31.6
Adse	1.3	0.3	76.7	130	150	64.5
Vendsava	1.4	1.7	50	211	98	25
Rinda	1.0	14.7	40.6	235	76	35
Spoleni	1.2	0.01	60.3	350	146	55
Gaiki	0.7	6.4	64.4	396	196	6
Kandava	1.3	18.4	27	394	80	19
Spara	0.6	7	50.8	264	87	86
Dundaga	1.0	4.7	40.6	228	124	52

The parameters of the first three layers were fixed at values which had been determined in the first stage of the solution. As a result of the interpretation of separate solutions, the value of the layer thickness h_4 which defines the depth of the conducting mantle surface, appeared to have a considerable scatter (see Table 1). In reality, the depth of the lower boundary of that layer must be approximately the same through the investigated area. For this reason we had left this parameter as a quantity to be found at the second stage too. Its initial value was taken in accordance with the condition $\sum_{i=1}^{4} h_i = 280$ km. Beginning from the depth of 280 km to 1,200 km, the formation was divided into 4 layers $h_5=107$ km, $h_6=263$ km, $h_7=550$ km, $h_8=\infty$. Introducing a greater number of layers appeared to have no important effect, since it increased the error in the determination of the parameters, whereas the agreement with the experimental data remained almost unchanged. The initial values of resistivities were chosen under the assumption that this quantity falls off exponentially with depth from 5,000 Ohm m to 0.1 Ohm m. To decrease the number of unknown parameters, the thicknesses of remaining layers were fixed and the values of resistivity ρ_5, ρ_6, ρ_7, and ρ_8 were estimated, which together with h_4 represent 5 components of the vector $\vec{\theta}$ denoting the unknown quantities in the algorithm.

The problem was solved using the algorithm of joint interpretation, POROKHOVA (1975). To obtain estimates for unknown parameters the maximization of the response function $l(\vec{\theta})$ was performed, with

$$l(\vec{\theta}) = -\sum_{m=1}^{M}\left\{K_m \ln(2\pi D_m)^{1/2} + \sum_{k=1}^{Km} \ln[\rho_T^m(\vec{\theta}, T_{k,m})] + \frac{1}{2D_m^2}\sum_{k=1}^{Km}\left[\frac{\rho_T^{m,k} - \rho_T^m(\vec{\theta}, T_{k,m})}{\rho_T^m(\vec{\theta}, T_{k,m})}\right]^2\right\} \quad (3)$$

where $m=1, 2, \ldots, M$, M is the number of sounding points, $k=1, 2, \ldots, K_m$, K_m is the number of experimental data in the m-th sounding point, D_m the root mean square deviation in the m-th sounding point, $\rho_T^m(\vec{\theta}, T_{k,m})$ is the direct problem solution computed from (1) and $\rho_T^{m,k}$ are experimental data ρ_T for the period $T_{k,m}$ in the m-th sounding point. A nonlinear system of equations $\partial l(\vec{\theta})/\partial\theta_s=0$ was solved using the Newton-Le Came-Marquardt method. First partial derivatives were calculated analytically. Second derivatives which depend on the random values $\rho_T^{m,k}$ were substituted by their average values

(Le Came's correction), which approximation eliminates laborious calculations from Newton's method (POROKHOVA, 1971). The estimates obtained for the parameters $\hat{\theta}$ have all the properties of Bayes estimates, that is they are asymptotically effective with their mathematical expectation equal to zero. The efficiency of such estimates is described completely by the matrix (the Fischer's error matrix)

$$||A_{ss'}|| = \left\| \sum_{m=1}^{M} \frac{1+2D_m^2}{D_m^2} \sum_{k=1}^{Km} \frac{\partial}{\partial \theta_s} \left[\ln \rho_T^m(\vec{\theta}, T_{k,m}) \right] \frac{\partial}{\partial \theta_{s'}} \left[\ln \rho_T^m(\vec{\theta}, T_{k,m}) \right] \right\|^{-1}_{\vec{\theta}=\hat{\theta}} \qquad (4)$$

where $s=1, 2, \ldots, S$ (S is the number of unknown parameters). The error is calculated from the expression $\gamma_s = \sqrt{A_{ss}}/\hat{\theta}_s$ where the numerator is the r.m.s. deviation of the estimate $\hat{\theta}_s$ from its average value. The correlation coefficients for the parameters θ_s and $\theta_{s'}$, are calculated by means of the normalization of the matrix $||A_{ss'}||$, so that $r_{ss'} = A_{ss'}/\sqrt{A_{ss}A_{s's'}}$.

If $r_{ss'}$ are small compared with unity, then the investigated experimental data set has a good resolving capability with respect to the parameters to be found, i.e. the influence of one parameter on the experimental field is different from that of another. A strong correlation means that corresponding parameters are unresolvable with the given data set, since the variations in the first group of parameters cancel out variations in the other group. The results of the interpretation are listed in Table 2, along with the r.m.s.

Table 2. Joint interpretation data. Thickness in km, resistivity in Ohm m, denominators of fractions give errors in percent. The thicknesses h_5, h_6, h_7 are fixed and equal in km: 107, 263, 550.

Model	l	Thickness of 4 layers	Resistivity in Ohm m			
			5	6	7	8
1. r_1=6,360 km	−1,570	235/8.5	5,955/162	25.3/21.6	1.14/20.8	0.04/195
1′. $r_1=\infty$	−1,571	234/14.2	4,795/970	24.3/25.3	1.04/22.3	0.02/177
2.	−1,584	190/22	903/141	63/25	1.93/27.6	0.04/116
3.	−1,623	80	1,000/99	35/16	1.19/45.8	0.085/257

relative errors of the estimated parameters. The obtained parameters differ remarkably from the initial ones, the resistivity of the fifth layer being changed the most considerably. Instead of a smooth transition from the highly resistive layer at the depth of 200–300 km to the layer with small resistivity at the depth of 400–600 km, as was assumed in the initial approximation, there appears to be an abrupt decrease of the resistivity from 5,000 Ohm m to 25 Ohm m. This sharp change, according to the interpretation, is located at the depth of 340 km. The computed distribution of conductivity is shown in Fig. 2, the initial approximation and the confidence intervals being indicated by dashed lines. The sounding curves plotted from the results of the interpretation reveal a satisfactory agreement with experimental points through the whole interval of periods (Fig. 1). Since the experimental data contain large errors, the solution of the inverse problem is not unique. We have found but one of the possible solutions.

In this respect we have studied the solution obtained with the assumption of another

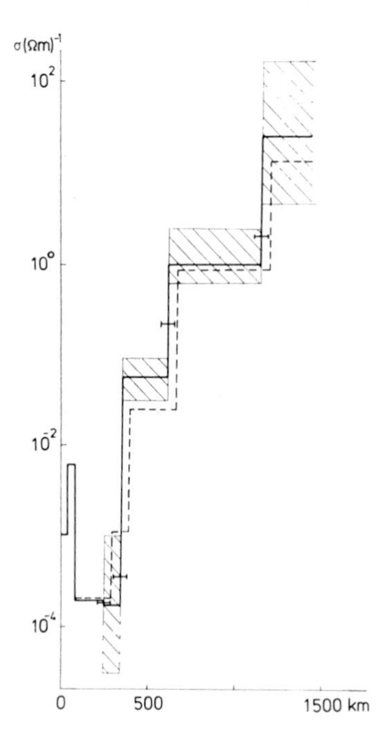

Fig. 2. Distribution of electrical conductivity with depth according to model 1. Shaded areas indicate confidence intervals for conductivity, cross bars show the confidence intervals for depths. The initial model is given by the dashed line.

initial distribution, in which the thickness of the fourth layer with the resistivity $\rho_4=$ 5,000 Ohm m was determined from the condition $\sum_{i=1}^{4} h_i=80$ km. Then the half-space was divided into four layers identical to those in model 1. As a result of the interpretation we have arrived at model 2, which reveals a smoother decrease of resistivity with depth. However the value of l appeared to be somewhat less than in the previous case (see Table 2). The model 3 appears to be even less credible; here, in contrast with model 2, the thickness of the fourth layer, h_4, was fixed by the relation $\sum_{i=1}^{4} h_i=80$ km. We have to point out one more interesting feature. As the computations have shown, the solution of the inverse problem can be performed in the framework of the flat Earth model with a good accuracy. Using in the interpretation algorithm the formula (1) for the calculation of the theoretical value of the apparent resistivity $\rho_T(\vec{\theta}, T)$, and taking $z_1=\infty$, we obtain a profile which is identical to model 1 (see model 1′, Table 2).

Let us compare the obtained distribution of the conductivity (model 1) with those in models suggested by other authors. Figure 3 shows the best known conductivity distribution curves obtained by means of interpretation of magnetic variation data only. The great difference in the proposed models may be explained by the fact that the inverse problem solution is not unique if there is a significant scatter in the input data. This non-uniqueness was aggrevated also by the fact that the magnetic variation data were restricted to the range between diurnal (small periods) and semiannual (long periods) variations. As a consequence, the values of conductivity remain unknown to depths of 200–300 km, and lower than 1,000 km. In order to lend a more quantitative sense to our speculations and comparisons, we have studied the efficiencv of our models of the conductivity distri-

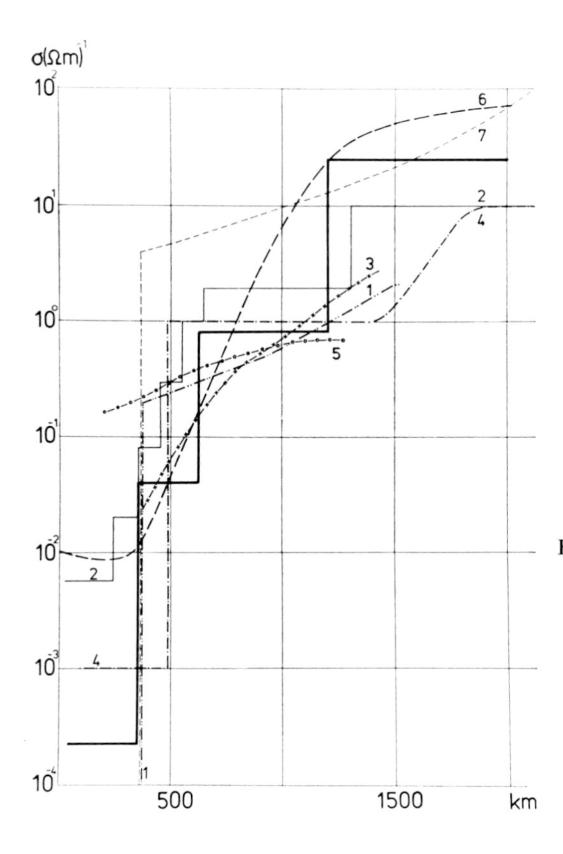

Fig. 3. Models of conductivity distribution. Legend for plots: 1—model of RIKITAKE (1966), 2—BANKS' (1972) model, 3—model of FAINBERG-ROTANOVA (1974), 4—model of DMITRIEV *et al.* (1977), 5—model of PARKER (1970), 6—model of MCDONALD (1957), 7—model of YUKUTAKE (1959). Thick solid line corresponds to model 1 of the present paper.

bution, from a depth of 250 km. Beginning from this depth the conductivity distribution in model 1 was replaced by that according to the above mentioned models and the quantity l was calculated. It can be treated as a measure of the agreement of models with our experimental data. The greatest values of l correspond to the models of RIKITAKE (1966) (curve 1 in Fig. 3) in which the conductivity jump is located at a depth of 374 km. The model of BANKS (1972) (curve 2 in Fig. 3) obtained as a result of the interpretation of the greatest amount of magnetic variations data, was shown to be slightly worse than the first two models. But the discrepancy may be removed to a great extent if the surface of the first conductivity jump is displaced from 250 km to 350 km in Banks' model. The value of l appears in that case (model 2′, in Table 3) to be close to those corresponding to model 1. The distributions of FAINBERG and ROTANOVA (1974) (curve 3 in Fig. 3) and DMITRIEV *et al.* (1977) (curve 4 in Fig. 3) are also consistent with our model if the conductivity distribution for the first 250 km is taken as that of model 1. The least satisfactory agreement with the magnetotelluric and the magnetic variations data is observed for PARKER's (1970) model (curve 5 in Fig. 3), in spite of the fact that the distribution of conductivity is also similar for the first 250 km to that of model 1. Finally we have to point out that the electrical conductivity distribution obtained as a result of the joint interpretation of MTS and MV data (model 1) not only describes the conductivity profile in the north-west of the Russian Platform, but may be regarded as a good

Table 3. Comparison of efficiencies of models in the framework of magnetotelluric and magnetovariational data. Numbers of models according to Fig. 3.

Model	1	2	2′	3	4	5
1	−1,588	−1,648	−1,602	−1,603	−1,616	−1,738

Table 4. Comparison of efficiencies of models in the framework of magnetovariational data only. The numbers of model according to Fig. 3.

Model	1	2	3	4	5	6	7	Model 1 of this work
1	−100	−110	−113	−106	−173	−108	−138	−103

approximation for the global distribution of this quantity. This may be proved by testing the credibility of the model using the magnetic variation data only. For this purpose the values of l were determined for different models using only magnetic variation data set. As in the previous case, every model was replaced by the multilayered profile, as detailed as possible. The corresponding values of l for each model are listed in Table 4. The values of l for most of models are distributed in the interval −100 to −110. The worst, in the sense of magnetic variation data, are for the models of PARKER (1970) ($l=-173$) and YUKUTAKE (1959) ($l=-138$), the best for the model of RIKITAKE (1966) ($l=-100$). The nearest to it appeared to be model 1 developed by joint interpretation of magnetovariational and MTS data ($l=-103$).

It may be concluded from the study described above, that model 1 provides the best approximation to our experimental data. However, one cannot draw a final and unambiguous conclusion that there exists a jump in conductivity by two orders of magnitude at the depth of 340 km in the investigated region. Model 1 has only a slight advantage over models 2 and 3. Small distortions of the sounding curve due to horizontal inhomogeneity of the sedimentary cover would be enough to produce noticeable changes in the result. It is necessary to carry out similar investigations in other regions of the northwest of the Russian Platform to improve the model obtained.

REFERENCES

BANKS, R. J., The overall conductivity distribution of the Earth, *J. Geomag. Geoelectr.*, **24**, 337–351, 1972.

BERDICHEVSKY, M. N., L. L. VANYAN, and E. B. FAINBERG, Theoretical Principles in using electromagnetic variations to study the electrical conductivity of the earth, *Geomag. Aeron.*, **9**, 465–467, 1969.

BERDICHEVSKY, M. N., V. I. DMITRIEV, I. A. YAKOVLEV, B. P. BABROV, Y. K. KONONOV, and D. A. VARLAMOV, Magnetotelluric sounding of the *H* horizontally inhomogeneous media, *Izv. USSR Acad. Sciences, Fizika. Zemli*, No. 1, 80–92, 1973.

BORISOVA, V. P., M. N. BERDICHEVSKY, N. M. ROTANOVA, E. B. FAINBERG, and T. N. TSCHEREVKO, New data on global magnetic variations sounding of the Earth, in *Research in geomagnetism, aeronomy, and solar physics. M.*, **30**, 194–204, 1974.

CAGNIARD, L., Basic theory of the magnetotelluric method of geophysical prospecting, *Geophysics*, **18**, 605–635, 1953.

DMITRIEV, V. I., N. M. ROTANOVA, O. K. ZAKHAROVA, and O. N. BALITSCHEVA, Geoelectrical and geothermical interpretation of results of deep magnetic variations sounding, *Geomagn. Aeron.*, **17**, 315–322, 1977.

FAINBERG, E. B. and N. M. ROTANOVA, Distribution of electrical conductivity and temperature inside the Earth from data on deep electromagnetic sounding, *Geomagn. Aeron.*, **14**, 709–714, 1974.

JODENTSCHYKOVA, A., Results of global sounding the Earth from data of spectrum of 27-days variations of geomagnetic field, *Geomagn. Aeron.*, **15**, 317–320, 1975.

KOVTUN, A. A., Induction studies in stable shield and platform areas, Review, Third workshop on Electromagnetic Induction in the Earth, Sopron, 1976.

KOVTUN, A. A. and L. N. POROKHOVA, Estimate of efficiency of MTS in the north-west of Russian Platform, Problems in geophysics: Leningrad Univ., *Phys. Geol. Science*, **24**, 266–291, 1974.

KOVTUN, A. A., N. D. TSCHITSCHERINA, A. A. LIPATOV, and L. N. POROKHOVA, Electrical parameters of crust and upper mantle in the western Latvia, Problems in geophysics: Leningrad Univ., *Phys. Geol. Science*, **18**, 73–78, 1965.

MCDONALD, K. L., Penetration of the geomagnetic secular field through a mantle with variable conductivity, *J. Geophys. Res.*, **62**, 117–141, 1957.

PARKER, R. L., The inverse problem of electrical conductivity in the mantle, *Geophys. J. R. Astr. Soc.*, **22**, 121–138, 1970.

POROKHOVA, L. N., Solution of inverse problem of electromagnetic sounding, in *Statistical Methods of Interpretation of Geophysical Observations*, pp. 66–74, Leningrad, 1971.

POROKHOVA, L. N., Joint interpretation of amplitude curves of MTS for the purpose of determination of Earth's parameters of great depths, Izvestia ANSSSR, *Fizika Zemli*, No. 5, 47–55, 1975.

RIKITAKE, T., *Electromagnetism and the Earth's Interior*, Elsevier, Amsterdam, 1966.

ROKITYANSKY, I. I., *A Survey of Anomalies of Electrical Conductivity by the Method of Magnetic Variations Profiles*, Naukova Dumka, Kiev, 1975.

SCHMUCKER, U. and Y. YANKOWSKY, Geomagnetic induction studies and the electrical state of the upper mantle, *Tectonophysics*, **13**, 133–256, 1972.

SOCHELNIKOV, V. V., On the conductivity determination of deep layers of the Earth, Izvestia ANSSR, *Fizika Zemli*, No. 7, 65–71, 1968.

VANYAN, L. L., M. N. BERDICHEVSKY, E. B. FAINBERG, and M. V. FISKINA, The study of the asthenosphere of the East European platform by electromagnetic sounding, *Phys. Planet. Inter.*, **14**, 1–2, 1977.

YUKUTAKE, T., Attenuation of geomagnetic secular variation through the conducting mantle of the Earth, *Bull. Earthq. Res. Inst. Tokyo Univ.*, **37**, 13–32, 1959.

Connection between the Electric Conductivity Increase due to Phase Transition and Heat Flow

A. ÁDÁM

Geodetic and Geophysical Research Institute of the Hungarian Academy of Sciences, Sopron, Hungary

(Received November 20, 1977; Revised January 31, 1978)

The deepest electric conductivity increase determined by electromagnetic induction methods can be attributed to the conductivity increase observed in the laboratory during rock phase transition. It can be called the Ultimate Conducting Increase (or UCL), since no further increase of conductivity has yet been detected at greater depths. A positive dP/dT gradient is characteristic of rock phase transition, where P is the pressure in BARS, T the temperature in °C.

The aim of the research reported here was to look for an indication of a positive dP/dT gradient in the data set available, i.e. whether greater depths of the UCL (or greater pressures) correspond to greater temperatures at the depth of the phase transition. The surface heat flow (q in HFU), the characteristic part of which is coming from the upper mantle, can be used as an indicator of the temperature. Since in the upper mantle of the territories with lowest heat flow ($q<1$ HFU) there is no partial melting, it can be assumed that the depth of 250–300 km of the conductivity increase determined on platforms and crystalline shields corresponds to the depth of the rock phase transition. Using these data, which are very critical for our interpretation, an empirical formula was calculated by least squares fit between the depth of the conductivity increase H_{UCL} (km) and the surface heat flow q (HFU): $H_{UCL}=16.3+292.5\,q$. This formula indicates a positive dP/dT gradient. On the basis of the data set, the average H_{UCL} is 420 km and the average heat flow 1.38 HFU. From this formula, taking $dP/dT=30$ BARS/°C, it can be concluded that the temperature at a depth of 300–400 km below the platform areas is about 1,000°C less than the average.

1. Introduction

One of the first results of geomagnetic induction studies was the detection of an increase of electric conductivity in the seismic C layer of the spherically symmetric Earth, attributed to phase transition of the rocks (e.g. LAHIRI and PRICE, 1939). Information about the regional, and even local, variations in the depth of this increase could only be collated as a result of the practical application of the magnetotelluric method in the 1960's. Data show considerable scatter due to the great depths involved (approximately several hundred km). Recently a part of the data has been, however, re-evaluated on the basis of inhomogeneous models in order to avoid the effects of near surface field distortions.

In order to check the geothermal effect on the formation of electrically conducting zones, ÁDÁM (1976b) determined the connection between terrestrial heat flow and the depths of different conducting zones in the crust and upper mantle using data from the monograph "Geoelectric and Geothermal Studies" (1976). Three zones could be investigated; the deepest or "ultimate" conductivity increase (: UCL) at a depth greater than 250–300 km may correspond to the phase transition of rocks (e.g. olivine→beta phase) according to the laboratory experiments of AKIMOTO and FUJISAWA (1965). For all three zones, the depths, in km, were approximated by a function of the form $h=h_0q^{-a}$ where q denotes the value of the terrestrial heat flow in HFU at the surface. For the UCL the experimental connection was the most uncertain. At the same time, as mentioned by ÁDÁM (1976b), it was contrary to the anticipated positive value of dP/dT for the phase transition of rocks (RINGWOOD, 1976). The ideas proposed then to explain the connection between the depth of the conductivity increase corresponding to the phase transition and the temperature (or—to a first approximation—the surface heat flow) must now be reconsidered.

The UCL data were re-examined in order to look for any indication of the existence of a positive gradient dP/dT. As an indicator of the deep temperature, only the surface heat flow can be used. Recent computations for the stationary case (POLLACK and CHAPMAN, 1977) have shown that the part of the heat flow characteristic of great tectonic units comes mainly from the upper mantle, therefore the thermic conditions there can be characterized by it. Depth data for the UCL were arranged according to regional heat flow values.

2. Revision of the Available Data Set

The data for the conductivity increase corresponding to rock phase transition (in short UCL data) were taken mainly from the previously mentioned monograph, taking into account KOVTUN's (1976) critical remarks on the great depth values in the NW part of the Russian Platform, i.e. in the Baltic Depression (Table 1). The reliability of the heat flow values is rather variable. In some cases, data are valid for the immediate vicinity of the MT sounding; in other cases, only regional averages or ranges are available (LUBIMOVA *et al.*, 1973). The UCL data themselves are of different kinds, some being individual soundings and others ranges from several measurements. All these data were taken into account with equal weight as a first approximation.

Figure 1 shows the depths of the UCL as a function of the heat flow, Fig. 1a showing the heat flow ranges and Fig. 1b average heat flows; the number of data for each depth interval can be seen in Fig. 1c. The data show considerable scatter. On platforms, and in some cases on crystalline shields, with very low heat flows ($q<1$ HFU), the soundings indicate the ultimate conductivity increase, or increases determined as ultimate ones, at two depth ranges, as seen on the left side of Figs. 1a and b. The greater depths ($H>$ 500 km) come from the NW part of the Russian Platform. KOVTUN and CHICHERINA (1976) determined the depth range to be 500–550 km allowing for an S-effect (i.e. the distortion effect of the near-surface inhomogeneities). In her review at the IAGA Workshop (Sopron, 1976), Kovtun questioned these depths as the E-polarisation curves showed a conductive layer at depths of 40–70 km. For a considerable part of the Russian or Easteuropean Platform she proposed a depth of 200–350 km for the UCL. Similar values

Table 1. Depth of conductivity increase and surface heat flow.

Tectonic unit or measuring point	Depth of the UCL [km]	Author	Heat flow HFU $=1\mu cal/cm^2 s$	Author	Remarks
Russian Platform**	200–350	Kovtun (1977)	0.8–1.2*	Lubimova *et al.* (1973)	*Heat flow range (Δq) **Several measuring points
Borok	300	Bryunelli *et al.* (1963)			
Voronezh Massif**	250–300	Anishchenko (1976)	0.8–1.2*	〃	
Siberian Platform** (Marginal parts)	270±40	Pospeev and Mikhalevsky (1976)	0.8–1.2*	〃	
Siberian Platform** Southeastern part (Aldan shield, etc.)	200–260	Berdichevsky *et al.* (1969)	0.8–1.2*	〃	
North-American Platform Western part	200–350	Porath and Gough (1971)	0.8–1.2		
Canadian shield** Southeastern part	170–240	Dowling (1970)	1?	World Data Center A (1976)	Terrestrial Heat Flow Map
South Australia**	230	Lilley and Tammemagi (1972)	0.8–1.2?		
Russian Platform** Northwestern part (Baltic depression)	500–550	Kovtun and Chicherina (1976)	0.8–1.2*	Lubimova *et al.* (1973)	
Bohemian Massif					
Mrákotin	350–400	Pěčová *et al.*		Čermák	
Hrádek	400–500	(1976)	1.2	(1976)	
Budkov	360				
Arctic Ocean	302	Trofimov			
Lomonosov-ridge	320	and	1.2	Lubimova *et al.*	
	320	Fonarev (1976)		(1976)	

Table 1. (cont'd)

Tectonic unit or measuring point	Depth of the UCL [km]	Author	Heat flow HFU =1 μcal/cm²s	Author	Remarks
Podvodniki Basin	325	(1976)	1.3		
Kara Kum Platform					
Northern part**	170–470	AVAGIMOV *et al.*	1.2	LUBIMOVA *et al.*	
Southern part**	500	(1976)	(0.8–1.6)*	(1973)	
Ural	490		1.4		
	520	KRASNOBAEVA	(1.2–1.6)	KUTAS *et al.*	
	520	*et al.* (1976)		(1976)	
	650				
Saxo-Thüringian					
block (Czechoslovakia)		PĚČOVÁ *et al.*		ČERMÁK	
Krupá	400–500	(1976)	1.4–1.6	(1976)	
Hawaii	330–380	HUTTON (1977)	~1.5	World Data	Terrestrial
Oahu Island		LARSEN (1975)		Center A	Heat Flow
				(1976)	Map
Little Plain	350–410	ÁDÁM and	1.5	BOLDIZSÁR	(Interpolation)
(Hungary)		VERÖ (1967)		(1964)	
Nagycenk					
Freiberg (GDR)	300	PORSTENDORFER	~1.9	HURTIG and	
		(1967)		SCHLOSSER	
				(1976)	
France					
Chambon la Forêt	750	FOURNIER *et al.*	q=	ČERMÁK	Not published
Garchy	800	(1971)	1.7–2–2.2–	(1976)	map
Plachez du Morvan	850		2.4–2.4–2.5		
Parc St. Maur	650				
Arctic Ocean		TROFIMOV and		LUBIMOVA *et al.*	
Haeckel-ridge	400	FONAREV	2.4	(1976)	
		(1976)			

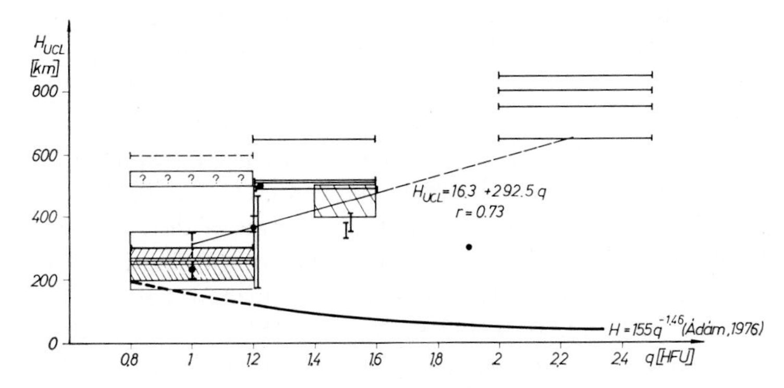

Fig. 1a. H_{UCL} data vs. heat flow. Hatched areas show the range of values in different tectonic units. The lower curve is the experimental depth vs. surface heat flow function for the conducting zone in the upper mantle (ÁDÁM, 1976b).

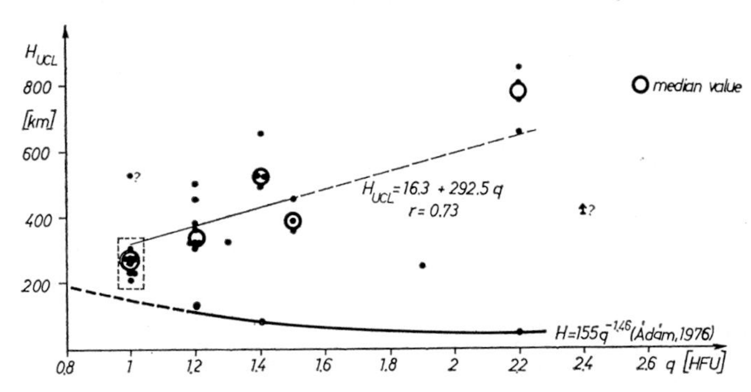

Fig. 1b. Similarly to Fig. 1a, but only averages are shown.

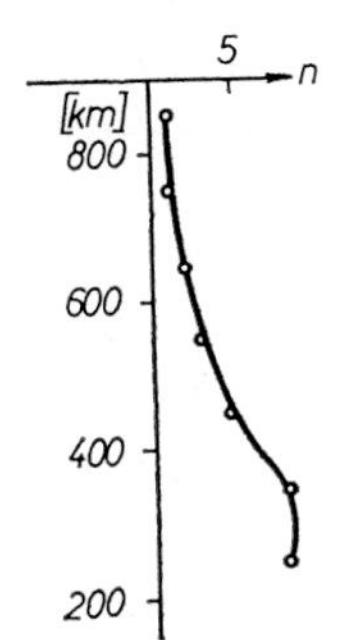

Fig. 1c. Number distribution of H_{UCL} vs. depth.

have been found in the Siberian Platform by POSPEEV and MIKHALEVSKY (1976), in the W part of the North American Platform by PORATH and GOUGH (1971), in the SE part of the Canadian Shield by DOWLING (1970) and on the Australian Craton by LILLEY and TAMMEMAGI (1972).

It thus seems likely that the depth of the "ultimate" conductivity increase detectable by MT soundings is about 250–300 km beneath the platforms or crystalline shields with the lowest heat flow values. This point is very critical for our interpretation.

Two explanations have been proposed for the conducting zones at depths of 250–300 km in areas with low heat flows ($q<1$ HFU). It can be the effect of rock phase transition corresponding to a positive gradient dP/dT, or of partial melting at the bottom of the lithospheric plate, as supposed by POSPEEV and MIKHALEVSKY (1976).

The following factors favour an interpretation in terms of the rock phase transition rather than one indicative of partial melting at these depths: (a) The depth of 250–300 km corresponding to 1 HFU is much greater than that determined for the same heat flow from Ádám's experimental function (ÁDÁM, 1976b): $H^{(\mathrm{km})}=155\,q^{-1.46}$. This function is valid for the depth of the conducting zone caused by partial melting at the top of the asthenosphere (see Fig. 1). (b) For the case of $q<1$ HFU, Pollack's temperature curves do not intersect the 0.85 *Tm* curve where *Tm* is the volatile free (refractory) periodotite melting curve (Fig. 2). Pollack notes, in a personal communication, that "very nearly the same results occur from the intersection of the temperature curves from 100% of the solidus of periodotite with volatiles, principally H_2O and CO_2 present. The latter case permits an interpretation directly in terms of partial melting".

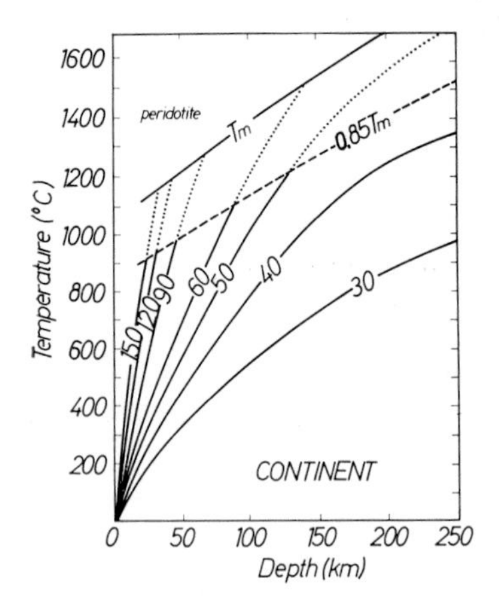

Fig. 2. Temperature vs. depth curves depending on heat flow in mWm^{-2} (1 HFU= 41.87 mWm^{-2}). *Tm* is the melting temperature of periodotite (POLLACK and CHAPMAN, 1977).

RINGWOOD (1962) also supposes that the Low Velocity Layer (LVL) corresponding to partial melting and so also to a conductive zone, may be absent at the bottom of the lithosphere due to the low temperature. These facts may indicate that the conductive zone at a depth of 250–300 km is connected with a phase transition.

The greatest depth values of about 800 km on the right side of Figs. 1a and b, are from France (FOURNIER *et al.*, 1971), from an area for which Cermák's heat flow map shows a heat flow of $q>2$ HFU.

The average of all depth values, for all heat flows, is 420 km (31 data).

3. Connection between the Depth of Conductivity Increase (H_{UCL}) and Surface Heat Flow

The linear least squares fit for the H_{UCL} versus q data yielded the equation: $H_{\mathrm{UCL}}=$

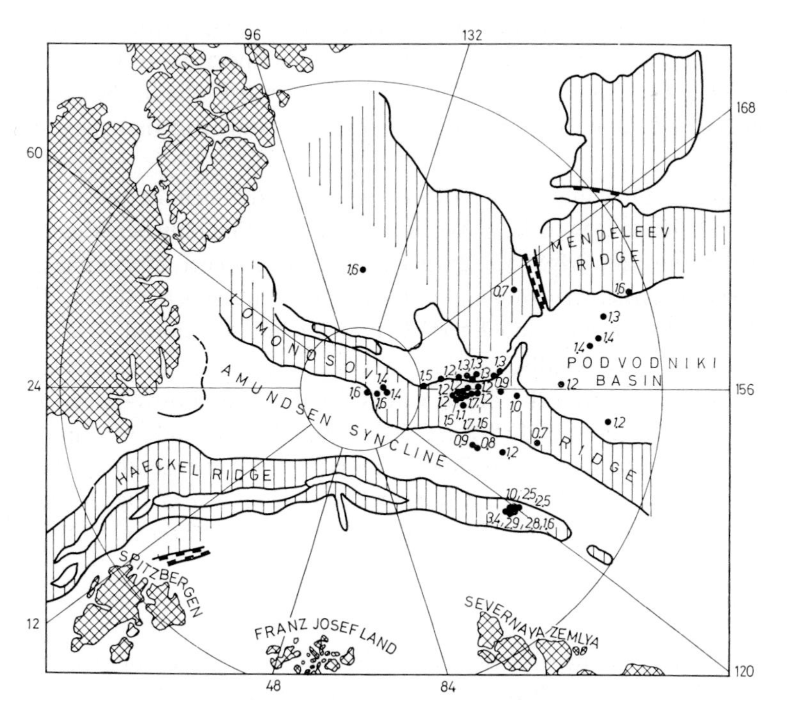

Fig. 3. Heat flow values in the Arctic Ocean (LUBIMOVA *et al.*, 1976).

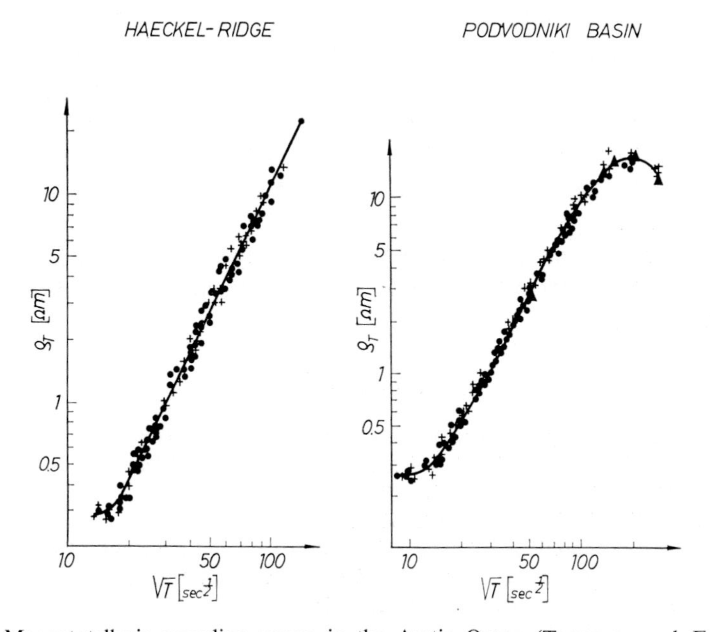

Fig. 4. Magnetotelluric sounding curves in the Arctic Ocean (TROFIMOV and FONAREV, 1976).

16.3+292.5 q. The correlation coefficient between H_{UCL} and q from 31 data is +0.73. As has been shown the depth increases with increasing temperature or heat flow indicating a positive dP/dT. RINGWOOD (1976) has shown an upward lobe in the phase transition surface of sinking slabs due to their lower temperature, similar to that suggested here for platform areas.

The very great depth values from France cannot be explained, despite conformity with the previously mentioned tendencies. We do not consider that only here there is an electric indication of the transition $MgSiO_4 \rightarrow 2MgO + SiO_2$ with absence of the phase transitions of the pyroxene into garnet structure and of the olivine into the beta phase. Even without these data, the trend between 1 and 1.5 HFU is as already mentioned.

The phenomenon was studied also in an area, where distortions are less probable. In the Northern Arctic Ocean the depth values for the Lomonosov-ridge (q=1.2 HFU) and Podvodniki Basin (q=1.3 HFU) are 302, 320, 320 km, and 325 km, respectively. On the Haeckel-ridge, the mean heat flow is 2.4 HFU, and the depth of the conductive zone is greater than about 400 km. The heat flows are shown in Fig. 3 after LUBIMOVA *et al.* (1976) and the MTS curves in Fig. 4 after TROFIMOV and FONAREV (1976).

4. Conclusions

A re-evaluation of the depth data of the UCL has yielded the following results:

1) On platform areas with the lowest heat flow values an "ultimate" conductivity increase was detected at depths of about 250–300 km. If we assume that this can be attributed to phase transition and not to partial melting, the linear least squares fit of H_{UCL} on q indicates a positive dP/dT.

2) The temperature at a depth of 300–400 km below platform areas is about 1,000°C less than the average, assuming a positive gradient dP/dT=30 BARS/°C.

3) If partial melting, nevertheless, occurs in a thin layer, then the resulting increase in conductivity cannot be distinguished from the conductivity increase resulting from phase transition and the former would increase the effect of the latter. This has been stated earlier (ÁDÁM, 1968).

As the data available are rather uncertain, this study has yielded only tentative results. A more accurate determination of the tendency apparent here can be made only with better data pairs H_{UCL} and q.

REFERENCES

ÁDÁM, A., Electric conductivity structure of the upper mantle in the Hungarian Basin, The problem and specialities of its determination, Thesis, Hangarian Academy of Sciences, Budapest, 1968.

ÁDÁM, A. (editor), *Geolectric and Geothermal Studies* (*East-Central Europe, Soviet-Asia*) *KAPG Geophysical Monograph*, Akadémiai Kiadó, Budapest, 1976a.

ÁDÁM, A., Quantitative connections between regional heat flow and the depth of conductive layers in the Earth's crust and upper mantle, *Acta Geod. Geophys. Mont. Hung.*, **11**, 503–509, 1976b.

ÁDÁM, A. and J. VERÖ, Latest results of electromagnetic measurements in Hungary, *Geofizikai Közlemények*, **16**, 25–52, 1967 (in Hungarian).

AKIMOTO, S. J. and H. FUJISAWA, Demonstration of the electrical conductivity jump produced by the olivin-spinel transition, *J. Geophys. Res.*, **70**, 443–449, 1965.

ANISHCHENKO, G. N., Deep magnetotelluric surveys in the central part of the Russian Platform, in *Geoelectric and Geothermal Studies*, edited by A. Ádám, pp. 628–634, Akadémiai Kiadó, Budapest, 1976.

AVAGIMOV, A. A., T. ASHIROV, V. G. DUBROVSKY, and K. NEPESOV, Deep magnetotelluric surveys in Turkmenia and Azerbaijan, in *Geoelectric and Geothermal Studies*, edited by A. Ádám, pp. 652–666, Akadémiai Kiadó, Budapest, 1976.

BERDICHEVSKY, M. N., V. P. BORISOVA, V. P. BUBNOV, L. L. VANYAN, I. S. FELDMAN, and I. A. YAKOVLEV, Anomaly of the earth's crust electroconductivity in Yakutiya, *Fizika Zemli*, **10**, 43–49, 1969 (in Russian).

BOLDIZSÁR, T., Terrestrial heat flow in the Carpathians, *J. Geophys. Res.*, **69**, 5269–5275, 1964.

BRYUNELLI, B. E., A. A. KOVTUN, O. M. RASPOPOV, N. M. RUDNEVA, and N. D. CHICHERINA, Magnetotelluric sounding in the Russian Platform in *Electromagnetic Sounding and Magnetotelluric Research Methods*, pp. 103–110, Izdatelstvo Leningradskogo Universiteta, Leningrad, 1963 (in Russian).

ČERMÁK, V., Heat flow investigation in Czekoslovakia, in *Geoelectric and Geothermal Studies*, edited by A. Ádám, pp. 414–425, Akadémiai Kiadó, Budapest, 1976.

ČERMÁK, V., J. PĚČOVÁ, and O. PRAUS, Heat flow, crustal temperatures and geoelectric cross-section in Czechoslovakia, in *Geoelectric and Geothermal Studies*, edited by A. Ádám, pp. 538–543, Akadémiai Kiadó, Budapest, 1976.

DOWLING, F. L., Magnetotelluric measurements across the Wisconsin arch, *J. Geophys. Res.*, **75**, 2683–2698, 1970.

FOURNIER, H. G., A. Ádám, L. De MIGUEL, and E. SANCLEMENT, Proposal for a first upper mantle magnetotelluric E-W profile across Europe, *Acta Geod. Geophys. Mont. Hung.*, **6**, 459–477, 1971.

HURTIG, E. and P. SCHLOSSER, Geothermal studies in the GDR and relation to the geological structure, in *Geoelectric and Geothermal Studies*, edited by A. Ádám, pp. 384–395, Akadémiai Kiadó, Budapest, 1976.

HUTTON, V. R. S., Induction studies in rifts and other active zones, *Acta Geod. Geophys. Mont. Hung.*, **11**, 347–376, 1976.

KOVTUN, A. A., Induction studies in stable shield and platform areas, *Acta Geod. Geophys. Mont. Hung.*, **11**, 333–347, 1976.

KOVTUN, A. A. and N. D. Chicherina, Deep magnetotelluric surveys in the central part of the Russian Platform, in *Geoelectric and Geothermal Studies*, edited by A. Ádám, pp. 620–628, Akadémiai Kiadó, Budapest, 1976.

KRASNOBAEVA, A. G., V. S. VISHNEV, and T. L. RUDNEVA, Deep electromagnetic studies in the Ural, in *Geoelectric and Geothermal Studies*, edited by A, Ádám, pp. 640–646, Akadémiai Kiadó, Budapest, 1976.

KUTAS, R. I., E. A. LUBIMOVA, and Ya. B. SMIRNOV, Heat flow map of the European part of the USSR and its geological and geophysical interpretation, in *Geoelectric and Geothermal Studies*, edited by A, Ádám, pp. 443–450, Akadémiai Kiadó, Budapest, 1976.

LAHIRI, B. N. and A. T. PRICE, Electromagnetic induction in nonuniform conductors, and determination of the conductivity of the Earth from terrestrial magnetic variations, *Phil. Trans. Roy. Soc.*, Ser. A **237**, 509–540, 1939.

LARSEN, J. C., Low frequency (0.1–6.0 cpd) electromagnetic study of deep mantle electrical conductivity beneath the Hawaiian Islands, *Geophys. J. R. Astr. Soc.*, **43**, 17–46, 1975.

LILLEY, F. E. M. and H. J. TAMMEMAGI, Magnetotelluric and geomagnetic depth sounding method compared, *Nature Phys. Sci.*, **240**, 184–187, 1972.

LUBIMOVA, E. A., V. N. NIKITINA, and G. A. TOMARA, *Heat Flow of the Inner and Marginal Sees of the USSR*, pp. 1–224, Nauka, Moscow, 1976 (in Russian).

LUBIMOVA, E. A., B. G. POLJAK, E. B. SMIRNOV, S. I. SERGIIENKO, E. B. KOPERBAKH, L. N. LIUSOVA, R. I. KUTAS, and F. V. FIRSOV, Review of heat flow data for the USSR, Heat flows from the crust and upper mantle of the Earth, in *Upper Mantle*, No. 12, pp. 154–195, Nauka, Moscow, 1973 (in Russian).

PĚČOVÁ, J., O. PRAUS, V. PETR, and M. TOBÝAŠOVÁ, Results of MT-soundings in the Bohemian Massif,

Geofysikální Sborník, **22**, 337–355, 1976.

POLLACK, H. N. and D. S. CHAPMAN, Mantle heat flow, *Earth Planet. Sci. Lett.*, **34**, 174–184, 1977.

PORATH, H and D. I. GOUGH, Mantle conductive structures in the Western United States from magnetometer array studies, *Geophys. J. R. Astr. Soc.*, **22**, 261–275, 1971.

PORSTENDORFER, G., Beiträge der Magnetotellurik zu einem komplexgeophysikalischen Nord-Süd-Profilschnitt durch die DDR, Report at the Geophysical Conference in Leipzig, 1967.

POSPEEV, V. I. and V. I. MIKHALEVSKY, Deep magnetotelluric surveys of the south of the Siberian Platform and in the Baikal rift zone, in *Geoelectric and Geothermal Studies*, pp. 673–682, edited by A, Ádám, Akadémiai Kiadó, Budapest, 1976.

RINGWOOD, A. E., A model for the upper mantle, *J. Geophys. Res.*, **67**, 857–867, 1962.

RINGWOOD, A. E., Phase transformations in descending plates and implications for mantle dynamics, *Tectonophysics*, **32**, 129–143, 1976.

TROFIMOV, I. L. and G. A. FONAREV, Deep magnetotelluric surveys in the Arctic Ocean, in *Geoelectric and Geothermal Studies*, edited by A. Ádám, pp. 712–716, Akadémiai Kiadó, Budapest, 1976.

Geomagnetic Variations Behavior in Central Europe

I. I. ROKITYANSKY

Institute of Geophysics, Kiev, U.S.S.R.

(Received December 10, 1977; Revised February 14, 1978)

The induction vectors and profile graphs of the anomalous field indicate clearly the existence of narrow elongated electrical conductivity anomalies in Central Europe. The anomaly map and parameters are presented. According to the analysis of synchronous variations in the region under study the horizontal magnetic field hodographs of the bay variations over the central part of the Carpathian anomaly are considerably stretched in the direction normal to the anomaly strike. The majority of permanent geomagnetic observatories of Central Europe are favourably located about the anomalies so that distortions of the horizontal magnetic field of bay variations are relatively small (except Niemegk and Moscow). The vertical component is distorted stronger. The behavior of electric component of bay and S_q variations over the central part of the Carpathian anomaly led us to suggest that this part of the anomaly connects conductors larger than itself. In all probability these are the North German-Polish anomaly and Black Sea.

In Central Europe the Wiese vectors (WIESE, 1965) have been generally used for presentation of anomalous field in vertical component of geomagnetic variations. The divergent change of the Wiese vectors direction indicates a location of the axis of a good conducting anomaly. The map of Wiese vectors (Fig. 1) and the results of synchronous profile observation present several large anomalies. The profile graphs of the anomalous field can be practically used only for estimation of the maximum possible depth of anomalous current centers (ROKITYANSKY, 1975). More exact determination of the anomalous body depth is important in discussing its nature and practical significance. The depth h of the upper surface of the well-conducting body can be determined by magnetotelluric sounding (MTS). The MTS-method is greatly influenced by distortion which makes it difficult to obtain reliable and precise results. Nevertheless magnetotelluric soundings performed over some anomalies allow us to make more concrete conclusions about the anomalous body upper surface. Results of interpretation of principal anomalies in Central Europe are given in Table 1.

North German-Polish (JANKOWSKI, 1965; PORSTENDORFER *et al.*, 1976; SCHMUCKER, 1959; and so on) and Dnieper-Donets (BONDARENKO *et al.*, 1973; ROKITYANSKY *et al.*, 1976) anomalies coincide with sedimentary basins of the same areas. MTS data give evidence of the near-surface nature of the anomalies. The resolution of the MTS-method does not allow us to answer the question of whether the anomalous conductor should be ascribed to the sedimentary rocks only, or whether it is partly localized in the basement rocks too.

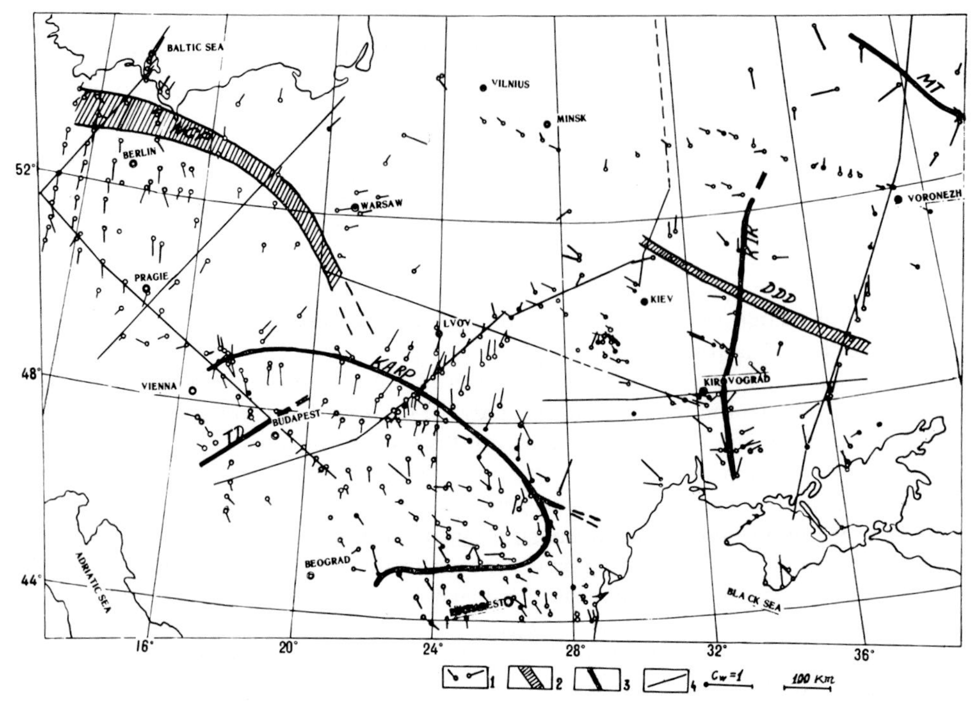

Fig. 1. Wiese vectors (1), principal electrical conductivity anomalies: mainly superficial, sedimentary basins (2), deep seated and intermediate (3), and regional deep seismic sounding profiles (4) in central Europe. Wiese vectors were drawn using the data: ÁDÁM *et al.*, 1972; BONDARENKO *et al.*, 1972, 1973; JANKOWSKI, 1965; LIPSKAYA *et al.*, 1972; MAKSIMOV *et al.*, 1976; PĚČOVÁ *et al.*, 1976; PORSTENDORFER *et al.*, 1976; ROKITYANSKY *et al.*, 1975, 1976, 1977; and WIESE, 1965.

The Carpathian anomaly is situated within the Alpine orogen area. MTS data indicate a deep location of the conducting body. Its probable nature is a partial melt at amphibole facies in the process of metamorphic transformation of sedimentary rocks accumulated in geosynclines (ROKITYANSKY, 1975).

The Transdanubian crustal anomaly was discovered by telluric currents and magnetotelluric sounding methods (ÁDÁM, 1976a). Its shallow depth (2–10 km according to more than 20 MTS measurements) excludes the partial melting explanation.

Kirovograd (ROKITYANSKY *et al.*, 1976) and Moscow-Tambov (MAKSIMOV *et al.*, 1976; ROKITYANSKY *et al.*, 1977) anomalies are located in the Precambrian basement of the East-European platform. Surface localisation of the Kirovograd anomaly is refuted by geological evidence, i.e., by the outcrop of high resistivity rocks of the Ukrainian shield. There were no reliable MTS results because of the complicated geology and the high level of industrial noise on the shield. MTS data received on the Moscow-Tambov anomaly indicate its crustal (not superficial) localisation. The anomaly has been traced out to 250 km only.

Fourteen selected geomagnetic bay and S_q-variations were analysed by the author and I.M. Logvinov using the records of 18 permanent European and 5 temporary Carpathian

Table 1. Electrical conductivity anomalies.

		Depth of upper edge h, km	Longitudinal conductivity G, $10^8\Omega^{-1}$m	Length km	Heat flow μcal/cm^{2}sec (Kutas *et al.*, 1976)	Geological situation
North-German-Polish	NGP	0	12	800	1.4–2	Elongated sedimentary basin.
Karpathian	KARP	15 ± 7	2	1,200	1.3–2	Alpine geosyncline. Likely nature of the anomaly is the deep sinking sediments partial melting occuring in the active amphibole stage of metamorphism.
Transdanubian	TD	6 ± 3	0.7	200?	1.8–2	Bacony Mountains of Paleozoic age.
Kirovograd	KIR	≤ 30	2	700	0.9–1.1	Precambrian basement the anomaly crosses Ukrainian Shield and Dnieper-Donetsk depression.
Dnieper-Donetsk	DDD	0	1.5	800	0.8–1.6	Elongated sedimentary basin.
Moskow-Tambov	MT	15 ± 10	2–3	250?	no data	Precambrian basement.

observatories. Synchronous hodographs of horizontal field are given on Figs. 2 and 3 for two bays of different polarisation. One can see that over the Carpathian anomaly (obs. TP, MG, OL, KR, RH) hodographs are 2–2.5 times longer than at the surrounding observatories KO, KV, HU, NG, and TH situated on the same latitude, and hodograph stretching is directed normally to the anomaly strike. For variations with periods $T > 60$ min the stretching diminishes to 1.5 for $T = 120$ min and to 1.2–1.3 for $T = 400$ min. Such dependence on period confirms the anomalous field frequency response of Carpathian anomaly observed earlier (ROKITYANSKY *et al.*, 1975).

The majority of the permanent geomagnetic observatories of central Europe are located outside the electrical conductivity anomalies, and the anomalous part of the horizontal magnetic field is relatively small for variations with periods more than 20 min. Small exceptions are the observatory Niemegk with an anomalous part for H of about 15% of normal (SCHMUCKER, 1959) and the observatory Moscow. They are situated near the south border of the North-German-Polish and Moscow-Tambov anomalies, respectively.

Variations of the vertical component Z at the observatories are less homogeneous because of a more gradual decrease of their anomalous part along the Earth's surface.

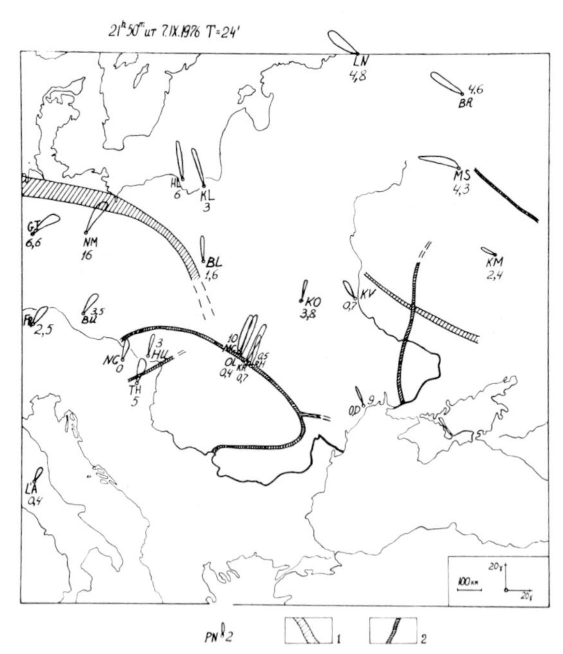

Fig. 2. Horizontal field polarisation for bay variation of Sept. 7, 1976, $21^h\ 50^m$ UT, $T = 24$ min. Permanent observatories: LN-Leningrad, BR-Borok, MS-Moscow, KL-Kaliningrad, HL-Hel, GT-Göttingen, NM-Niemegk, KM-Komsomolsky, BL-Belsk, KO-Korets, KV-Kiev, FR-Furstenfeldbruch, BU-Budkov, HU-Hurbanovo, NG-Nagycenk, TH-Tihany, OD-Odessa, LA-L'Aquila, PN-Pendeli. Temporary field observatories: TR, MG, OL, KR, RH. The figure near each observatory is the double amplitude of the vertical component, in nT. Electrical conductivity anomalies: 1, superficial and 2, deep seated and intermediate.

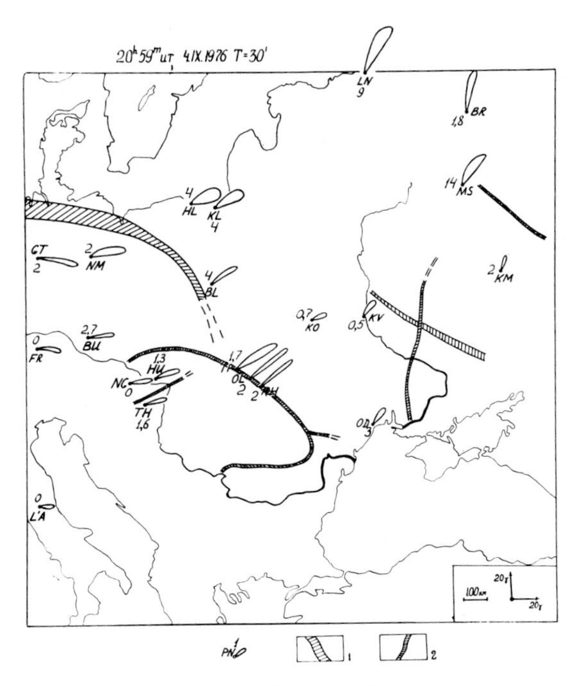

Fig. 3. Horizontal field polarisation for bay variation of Sept. 4, 1976, 20^{h} 59^{m} UT, $T=30$ min. Notation as for Fig. 2.

Amplitudes of Z given of Figs. 2 and 3 demonstrate the inhomogeneity and its dependence on horizontal field polarisation relative the nearest anomaly strike or nearest slope of well conducting sediments. Field observations TP, OL, KR, RH are situated near the axis of the Carpathian anomaly and Z-variations are small there. MG is situated 20 km to the north of the axis in the zone of maximum Z-variations.

Behavior of the horizontal electric field of bay and S_q variations is rather unusual in the Soviet Carpathians, i.e., in the central part of the anomaly. It is well known that longitudinal (and also transverse) electric (telluric) field in and over a two-dimensional good conductor is less than outside of it, and the difference diminishes to zero as period increases to direct current. Synchronous records of bay and S_q-variations have the opposite behavior (BONDARENKO *et al.*, 1972): over the Carpathian anomaly the electric field increases by several times. In 1973 and 1976 deep MTS observations were made at four points (OL, KR, RH_1, and RH_2, in Fig. 3) over the Carpathian anomaly. All the MTS-curves have descending parts for periods $T \leq 100$ sec, which confirms the existence of an electrical conductivity anomaly and determines its depth at 15 ± 7 km. A typical longitudinal MTS-curve (ROKITYANSKY *et al.*, 1975) is given on Fig. 4. For periods $T > 300$ sec we can see an unusually steep and long abcending branch of the MTS curve. Formal interpretation of the curve gives unrealistic parameters of deep conductivity distribution, with the well conducting basement at a depth of more than 1,000 km, which contradicts MTS data in the Pannonian basin (ÁDÁM, 1976a) and modern geophysical ideas. Following the hypothesis of constancy of deep ($h > 300$ km) geoelectric cross-section under all

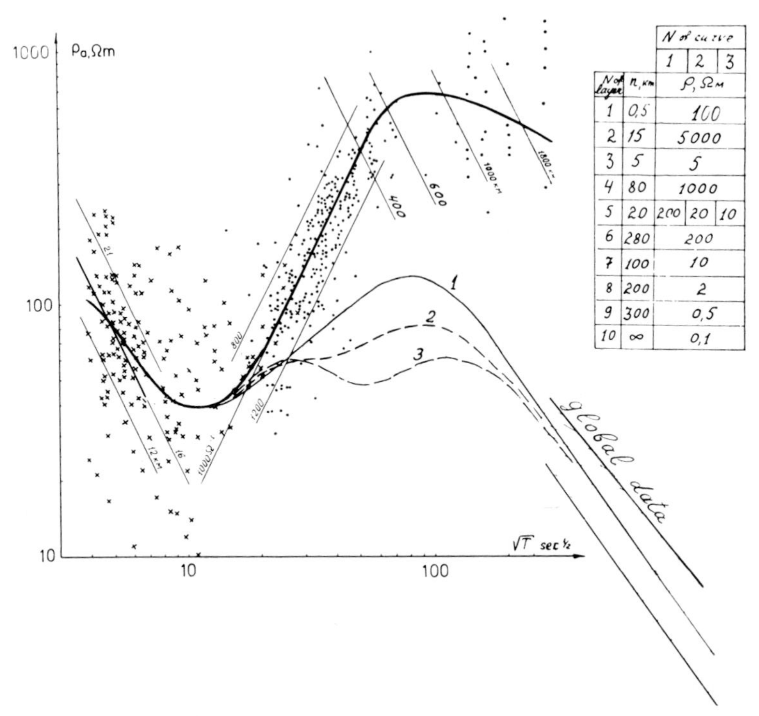

Fig. 4. MTS longitudinal curve (azimuth N 60 W) over Carpathian anomaly axis at RH. Crosses are results of observations at rapid-run stations, points are from standard records, 20 mm/h. Solid curve is the most probable one according to observed data. Tabulated, 10-layer model curves 1, 2 and 3 have fifth layer resistivity $\rho_5=200$, 20, and 10 Ωm, respectively.

the continents, and using global sounding data (ROKITYANSKY, 1970, 1975), the theoretical curves 1, 2, 3 (Fig. 4) were calculated. The long period part ($T>10^4$ sec) of the curves corresponds to global data and the short period ($T\leq 100$ sec) to all the MTS data observed over the Carpathian anomaly. The middle parts of the curves depends on longitudinal conductivity $S=h/\rho$ of the asthenosphere: curve 1 for $S=100\ \Omega^{-1}$ (without asthenosphere), curve 2 for $S=1{,}000\ \Omega^{-1}$, curve 3 for $S=2{,}000\ \Omega^{-1}$. Such values of S are infered for the East-European platform. Under the Pannonian basin and the Carpathians S may be a little more, so the MTS curve would be lower than curve 3 and the divergence from experimental data would be still greater.

The increase of the electric field over the anomaly and the consequent divergence of experimental and model MTS curves for periods more than 300 sec we can explain only by a sharp superchannelling of anomalous currents, which exceeds the concentration possible in a two-dimensional conductor. Therefore, the central part of the Carpathian anomaly should connect two conductors of larger conductivities than it has itself (ROKITYANSKY *et al.*, 1975). Such conductors may be the North German-Polish anomaly and the Black Sea (Fig. 5). Special observations of 1977 confirm the existence of the Carpathian anomaly branch directed to the Black Sea (Fig. 1). This effect of superchannelling of telluric currents should occur in narrow parts of the conductor (in sea straits, for instance),

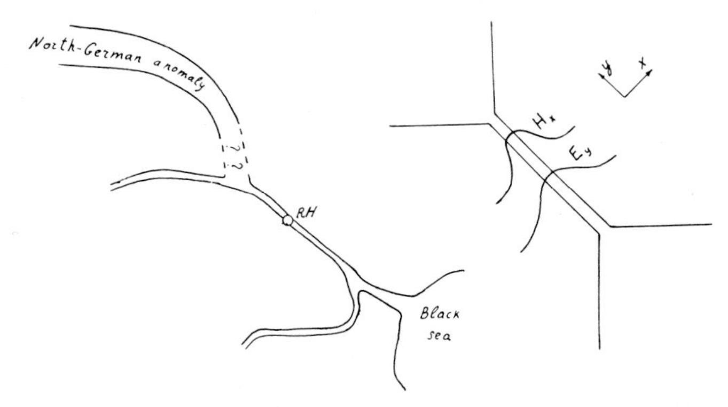

Fig. 5. Scheme of connection of the Carpathian anomaly (central part) to North German-Polish anomaly and to Black Sea (a), and scheme of the strait effect (b).

and in poor conducting parts which connect well conducting bodies (in a narrow isthmus between sea bays, for example). The reality of the superchannelling effect (or strait effect) is confirmed by physical modelling. Figure 5b illustrates the behavior of magnetotelluric field in this case.

My thanks are due to my colleagues I. M. Logvinov and S. N. Kulik for assistance in fulfilment of the work, and to all collaborators of geomagnetic observatories who have sent copies of magnetograms.

REFERENCES

Ádám, A., Results of deep electromagnetic investigations, in *Geoelectric and Geothermal Studies, KAPG Geophysical Monograph*, edited by A. Ádám, pp. 547–560, Akadémiai Kiadó, Budapest, 1976a.

Ádám, A., Distribution of electric conductivity in seismic (deep) fractures in Transdanubia, *Acta Geodaet., Geophys. et Montanist. Acad. Sci. Hung.*, **11**, 277–285, 1976b.

Ádám, A., J. Vero, and A. Wallner Regional properties of geomagnetic induction arrows in Europe, *Acta Geodaet., Geophys. et Montanist. Acad. Sci. Hung.*, **7**, 251–287, 1972.

Bondarenko, A. P., A. I. Bilinsky, and F. I. Sedova, *Geoelectromagnetic Variations in the Soviet Carpathians*, 116 pp., Naukova Dumka, Kiev, 1972 (in Russian).

Bondarenko, A. P., A. I. Bilinsky, and A. M. Shilova, On character of geomagnetic field bay variations in the Dnieper-Donets depression, *Geophys. Comm.*, **52**, 32–35, Naukova Dumka, Kiev, 1973 (in Russian).

Jankowski, J., Short-period variations of the Earth's magnetic field on the territory of Poland and their relations to deep substratum structure, *Acta Geophys. Polonica*, **13**, 1965.

Kutas, R. I., E. A. Lubimova, and Ya. B. Smirnov, Heat flow map of the European part of the USSR and its geological and geophysical interpretation, in *Geoelectric and Geothermal Studies, KAPG Geophysical monograph*, edited by A. Ádám, pp. 443–449, Akadémiai Kiadó, Budapest, 1976.

Lipskaya, N. V., M. S. Babushnikov, N. P. Vladimirov, N. A. Deniskin, M. K. Kravtsova, U. N. Kuznetsov, N. N. Nikiforova, and J. P. Khotko, Variations of natural electromagnetic field and electrical conductivity of the Earth interior, 157 pp., Nauka and Tekhnika, Minsk, 1972 (in Russian).

Maksimov, V. M., L. L. Van'an, and E. P. Kharin, Magnetovariation anomaly within the Voronezh crystalline massif, in *Geomagnetic Researches N. 15*, pp. 90–102, Sov. Radio, Moscow, 1976 (in Russian).

Pěčová, J., V. Petr, and O. Praus, Depth distribution of the electric conductivity in Czechoslovakia from electromagnetic studies in *Geoelectric and Geothermal Studies, KAPG Geophysical Monograph*, edited by A. Ádám, pp. 517–537, Akadémiai Kiadó, Budapest, 1976.

PORSTENDORFER, G., W. GÖTHE, K. LENGNING, Ch. OELSNER, R. TANZER, and E. RITTER, Nature and possible causes of the anomalous behaviour of electric conductivity in the north of the GDR, Poland and the FRG, in *Geoelectric and Geothermal Studies, KAPG Geophysical Monograph*, edited by A. Ádám, pp. 487–500, Akadémiai Kiadó, Budapest, 1976.

ROKITYANSKY, I. I., Investigations of abyssal electroconductivity, *Geophys. Comm.*, **38**, 102–106, Naukova Dumka, Kiev, 1970 (in Russian).

ROKITYANSKY, I. I., Investigation of the electrical conductivity anomalies by the method of magnetovariation profiling method 279 pp., Naukova Dumka, Kiev, 1975 (in Russian).

ROKITYANSKY, I. I., S. N. KULIK, I. M. LOGVINOV, and V. N. SHUMAN, The electrical conductivity anomaly in the Carpathians, *Acta Geodaet., Geophys. et Montanist. Acad. Sci. Hung.*, **10**, 277–286, 1975.

ROKITYANSKY, I. I., S. N. KULIK, I. M. LOGVINOV, and V. N. SHUMAN, Deep magnetovariation studies in Ukraine, in *Geoelectric and Geothermal Studies, KAPG Geophysical Monograph*, edited by A. Ádám, pp. 634–639, Akadémiai Kiadó, Budapest, 1976.

ROKITYANSKY, I. I., I. M. LOGVINOV, and V. M. MAKSIMOV, Magnetometer array study in the central part of the Russian Platform, *Acta Geodaet., Geophys. et Montanist. Acad. Sci. Hung.*, **12**, 131–137, 1977.

SCHMUCKER, U., Erdmagnetische Tiefensondierung in Deutschlahd 1957–1959, *Abh. Acad. Wiss. Gottingen, Math. -Phys. kl. V. 1, H. 5*, 51 pp., 1959.

WIESE, H., Geomagnetishe Tiefentellurik, *Deutsche Acad. Wiss. Geomagn. inst. Potsdam*, Berlin, Abhandlung N36, 1965.

Geomagnetic Sounding of an Ancient Plate Margin in the Canadian Appalachians

J. A. WRIGHT and N. A. COCHRANE

Department of Physics, Memorial University of Newfoundland, St. John's, Canada

(Received January 8, 1978; Revised February 22, 1978)

A geomagnetic induction anomaly in eastern Newfoundland is identified with the line of closure of the proto-Atlantic Ocean. A model consisting of an electrically conductive fossil lithospheric slab embedded within the continental lithosphere is shown to be quantitatively consistent with experimental observations. The conductivity of the Appalachian central mobile belt in Newfoundland does not greatly exceed that of other crustal sections in Eastern Canada.

1. Introduction

The long period electromagnetic response of Newfoundland in the vertical field component has been mapped between geomagnetic latitudes 58°N and 63°N (COCHRANE and WRIGHT, 1977). The previous work, tentatively postulated, an electrical conductivity anomaly in eastern Newfoundland between the stations LHE and NWI (Fig. 1). With the additional analysis of a short stretch of data (27 hr) from NWG this conjecture becomes more probable. Figures 2 and 3 show the contoured in-phase and quadrature transfer function amplitudes at periods of 2,000 s and 1,000 s respectively.

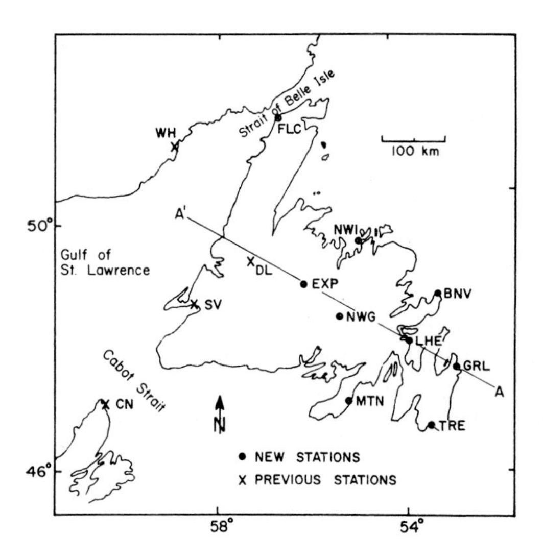

Fig. 1. Area map, showing station locations and profile AA′ through the Island of Newfoundland.

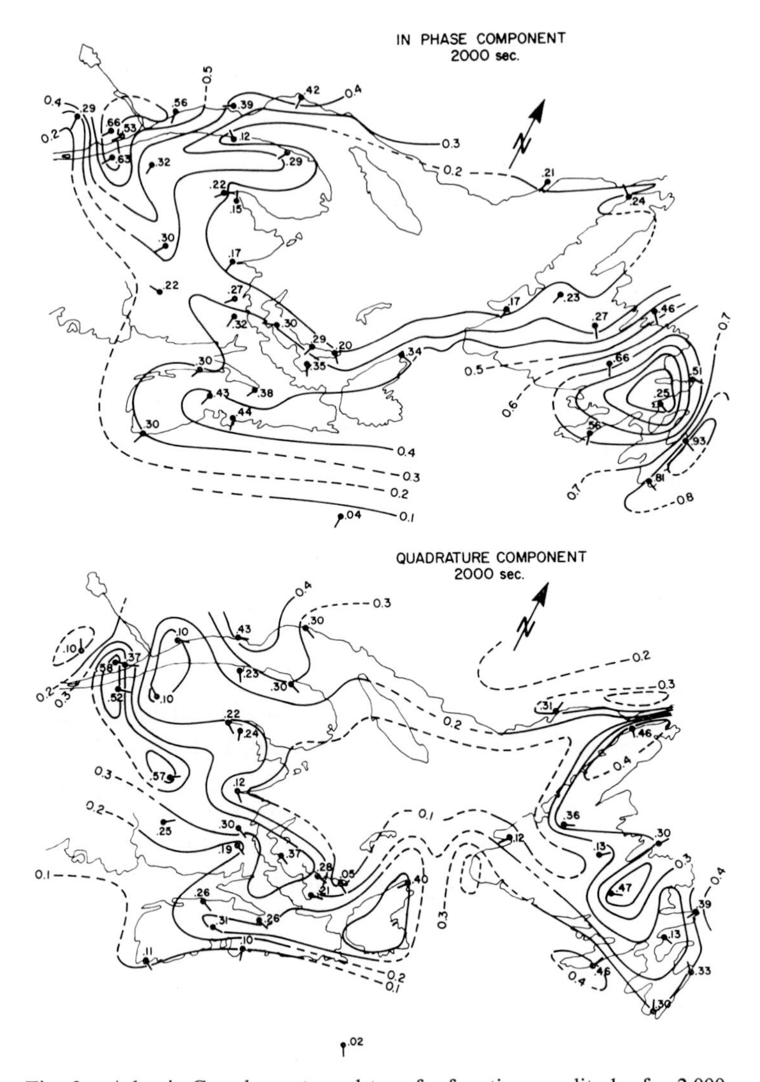

Fig. 2. Atlantic Canada contoured transfer function amplitudes for 2,000 s.

The anomaly is reflected in each plot as a semilinear feature characterized by relatively low transfer function amplitudes near LHE compared to amplitudes to the immediate southeast and northwest. The simplest interpretation consistent with the geometry and spatial extent of the anomaly would call for a good conductor to strike northeastward at a depth less than 100 km between LHE and the stations NWI and NWG to the west. The very high transfer function amplitudes east of LHE are attributed to coast effect due to the deflection of electric currents in the shallow water and sediments of the Grand Banks about the comparatively resistive rocks of the Avalon Peninsula of southeastern Newfoundland.

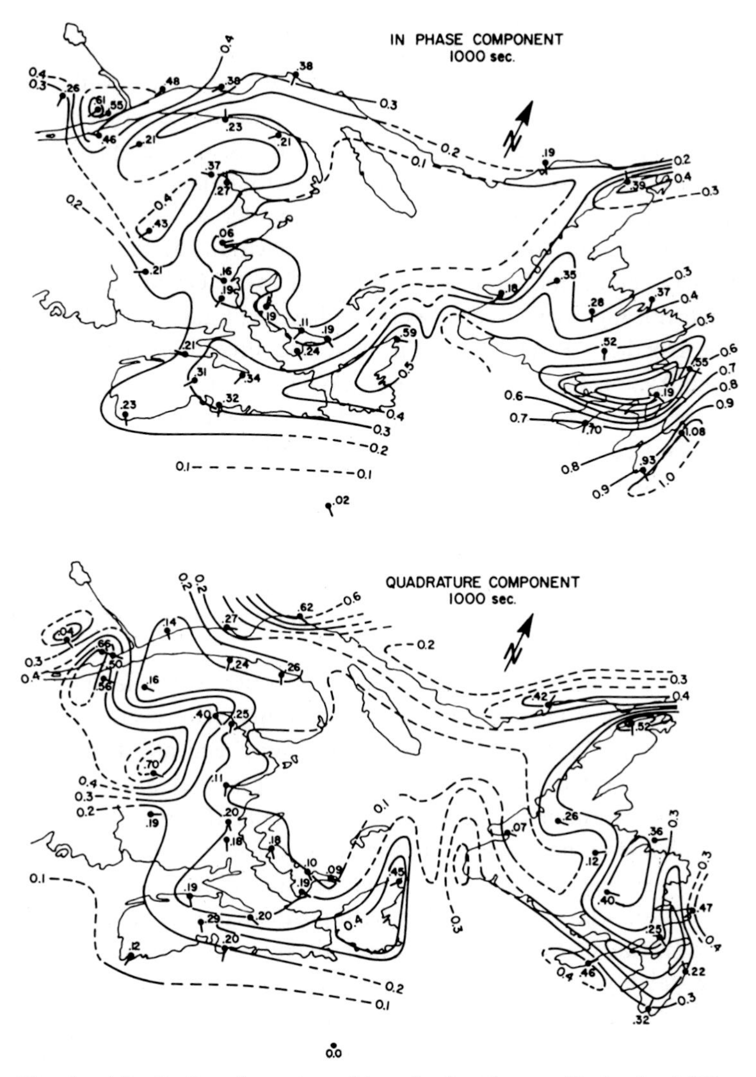

Fig. 3. Atlantic Canada contoured transfer function amplitudes for 1,000 s.

2. Geological Interpretation

The relative tectonic quiessence of the eastern margin of North America since the mid-Jurassic (RODGERS, 1970) rules out electrical conductivity anomalies arising from the abnormal lithospheric temperature distributions characteristic of active orogeny. However, pronounced conductivity anomalies have been observed in rocks as old as Precambrian (CAMFIELD and GOUGH, 1977; ROKITYANSKIY and LOGVINOV, 1972). LAW and RIDDIHOUGH (1971) have suggested conductivity anomalies in stable areas to be associated with geologic boundaries marking ancient plate margins. There is ample evidence such a boundary exists in the vicinity of the eastern Newfoundland conductor (STRONG *et al.*,

1974). CAMFIELD and GOUGH (1977) suggest several conductivity enhancement mechanisms of the well studied North American Central Plains anomaly, an apparent crustal suture type anomaly. Among these are conductive graphites associated with extensive shear zones and the probable incorporation of hydrated conductive oceanic materials into the lithosphere at the boundary. In the absence of evidence of widespread graphite zones or other superficial good conductors in eastern Newfoundland, the latter mechanism is developed and tested to determine whether it is compatible with our observations.

A prominant feature on the geological map of Newfoundland (WILLIAMS *et al.*, 1974) is the Appalachian central mobile belt consisting of lower Ordovician to mid-Devonian igneous and metavolcanic rocks wedged between basement gneisses and granites of Grenville age in the Northern Peninsula to the west and Precambrian volcanics and metasediments of the Avalon Platform to the east (WILLIAMS, 1964; WILLIAMS and STEVENS, 1974). Along the eastern margin of the central mobile belt lies a region of granites and gneisses of mainly Ordovician age fringed on the west by scattered ultramaties. This so-called Gander Zone is identified by WILLIAMS and STEVENS (1974) with an ancient continental margin now forming the eastern boundary of the Appalachian system. Geochemical evidence suggests the granites of this area to have been emplaced over an eastward dipping descending lithospheric slab (STRONG *et al.*, 1974) related to the closure of the "Iapetus Ocean" or proto-Atlantic which was finally complete by the mid-Devonian (MCKERROW and COCKS, 1977).

Most acidic continental type rocks are quite resistive at the temperatures expected in the crust and uppermost mantle (PARKHOMENKO, 1967). However, there are reasons to believe that serpentinites and amphibolites containing chemically bound water become quite conductive at elevated temperatures (A. Duba, personal communication). These materials should be common in the oceanic lithosphere (CHRISTENSEN and SALISBURY, 1975). If past subduction has occurred at the ancient plate margin in eastern Newfoundland a remnant descending lithospheric slab may remain embedded in the crust and upper mantle. Such a fossil lithospheric slab has been postulated under Nevada on the basis of seismic evidence (KOIZUMI *et al.*, 1973). Phase diagrams for amphibole (WYLLIE, 1967) indicate that the stability field extends to temperatures of 400°C at 20 Kbars. The present geotherm for Newfoundland is slightly subnormal (J. Wright, unpublished data), hence amphibole stability would extend to a maximum depth of burial of about 90 km. At greater depths any amphibole in the upper portion of the descending slab would have lost its chemically bound water reverting to stable and presumably much less conductive pyroxenes as the downthrust slab slowly came to thermal equilibrium. An electrical conductivity of 0.1 Siemens/m for amphibole would not be unreasonable (A. Duba, personal communication) on theoretical grounds although direct experimental measurements at deep crustal temperature and pressures are lacking. An important factor in determining the conductivity of any amphibolite facies rock would be the state of connectedness of the conductive mineral grains. A small change in amphibole abundance could profoundly alter the conductivity.

3. Quantitative Analysis

Before calculating the anticipated geomagnetic anomaly arising from the above model it will be essential to separate the signature of the inland anomaly in our experimental data from that of the superimposed coast effect. The coast effect in eastern Newfoundland and especially on the Avalon Peninsula is three dimensional. Reference to bathymetric charts show the Grand Banks protruding some 200 km southeast from the Avalon Peninsula to form a relatively narrow shallow water barrier separating the truly oceanic Labrador Sea from the comparable deep waters south of the Scotian Shelf. Because of the geometry of the Grand Banks barrier one might expect open ocean electric currents transiting between these basins on encountering the continental shelf to be concentrated within the thin water layer until they eventually reenter the deeper ocean. The process becomes one of simple DC type conduction within the shelf waters, the AC aspects of the process being controlled by the electromagnetic responses of the connected and much more extensive ocean basins. An analogy can be drawn with the well known case of telluric currents within sedimentary basins being deflected upward and concentrated above embedded salt diapirs.

With this simple model one can account for the anomalously large coast effect at GRL with surprising accuracy. Assuming current conservation to apply for electric currents transiting parallel to the coast, the expected coast effect at GRL should approximate

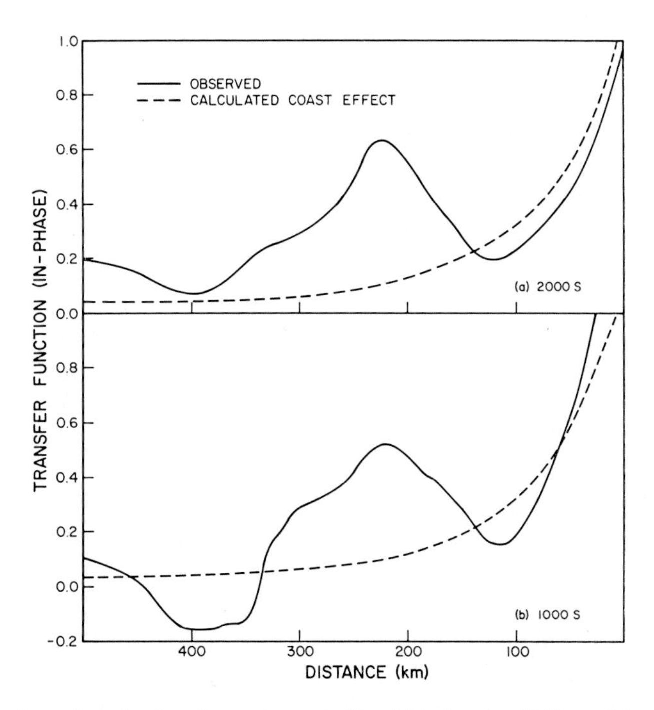

Fig. 4. In-phase transfer functions along profile AA′ showing 2-D model coast effect. a) 2,000 s; b) 1,000 s. Distances are measured northwestward along AA′ with the origin at the coast near GRL.

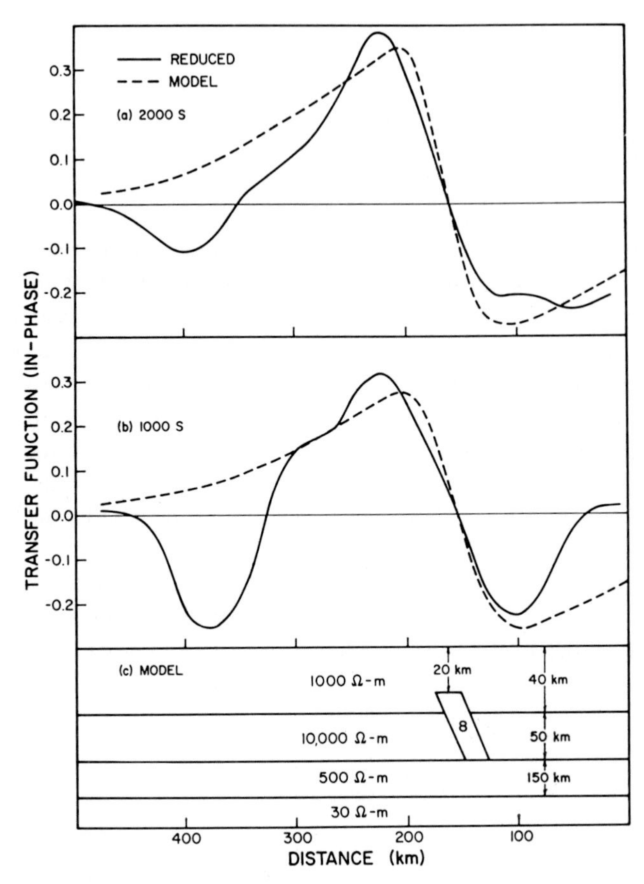

Fig. 5. Reduced in-phase transfer functions along profile AA′ showing 2-D model results for lithospheric slab. a) 2,000 s; b) 1,000 s; c) model properties. Profile origin as in Fig. 4.

that from a two dimensional 5 km deep ocean beginning about 10 km offshore to allow for the irregular nature of the coastline. Such a model was simulated on a 30×30 grid assuming an underlying Cantwell-MacDonald conductivity distribution. Very good agreement was obtained for both the in-phase and quadrature amplitudes within 100 km of the coastline.

Because of the relative complexity and uncertainty of the quadrature response far inland evident in Figs. 2 and 3, only the in-phase components are modelled. Figures 4a and b show in-phase single station transfer function amplitudes resolved along the profile AA′ (Fig. 1) at periods of 2,000 s and 1,000 s respectively. Also shown are the calculated coast effect amplitudes along the same profile. The differences in in-phase components are displayed in Figs. 5a and b. As a first order approximation the electromagnetic coupling between the coast effect and the inland conductor is ignored, allowing modelling of difference transfer functions. A model compatible with the above conjecture concerning subduction is shown in Fig. 5c. The conductive upper portion of a lithospheric slab is modelled as a series of 10 km square blocks descending at 45° to the east in the depth range 20 to 90 km. In-phase single station transfer functions numerically modelled on a 30×30 grid (WRIGHT, 1970) are displayed in Figs. 5a and b. A dipping conductive slab will give rise to a asymetrical anomaly, but the confidence limits on the experimental data do not permit reliable determination of the true direction of dip. To be fair one must point out that our data at present do not rule out other models such as a very conductive lower crust localized under the Gander Zone. However, the proposed model seems most compatible with the geological constraints. A denser network of stations recording at higher frequencies (thereby minimizing the coast effect) could resolve the direction of the dip.

4. Relation to Other Geophysical Features

At present it is not known if the proposed conductor might be an extension of the conductive region under the Scotian Shelf to the southwest (COCHRANE and HYNDMAN, 1974) or whether more likely the latter feature extends northeastward along the continental margin to underlie the Grand Banks. There is little evidence for a broad area of abnormally high conductivity under the central mobile belt in Newfoundland manifested in a region of low transfer function amplitudes as occurs in Virginia (EDWARDS and GREENHOUSE, 1975) or the western Cordillera (COCHRANE and HYNDMAN, 1970; CANER *et al.*, 1971). If the conductive region is restricted to the Gander Zone geological inference is that its westward extension lies through north-central New Brunswick (RAST *et al.*, 1976). Northeastward from Newfoundland the Gander Zone can be traced by other geophysical evidence as a narrow linear feature to the vicinity of the continental margin (JACOBI and KRISTOFFERSEN, 1976). It is a matter for speculation if the extensions or equivalents of the Gander Zone might be involved in the geomagnetic induction anomalies observed in central New Brunswick (COCHRANE and HYNDMAN, 1974) and the Caledonian system of Scotland (HUTTON *et al.*, 1977).

This work was supported by the NRC of Canada through grant A 7505 and by Memorial University of Newfoundland.

REFERENCES

CAMFIELD, P. A. and D. I. GOUGH, A possible Proterozoic plate boundary in North America, *Can. J. Earth Sci.*, **14**, 1229–1238, 1977.

CANER, B., D. R. AULD, H. DRAGERT, and P. A. CAMFIELD, Geomagnetic depth-sounding and crustal structure in western Canada, *J. Geophys. Res.*, **76**, 7181–7201, 1971.

CHRISTENSEN, N. I. and M. H. SALISBURY, Stucture and constitution of the lower oceanic crust, *Rev. Geophys. Space Phys.*, **13**, 57–86, 1975.

COCHRANE, N. A. and J. A. WRIGHT, Geomagnetic sounding near the northern termination of the Appalachian system, *Can. J. Earth Sci.*, **14**, 2858–2864, 1977.

COCHRANE, N. A. and R. D. HYNDMAN, A new analysis of geomagnetic depth-sounding data from western Canada, *Can. J. Earth Sci.*, **7**, 1208–1218, 1970.

COCHRANE, N. A. and R. D. HYNDMAN, Magnetotelluric and magnetovariational studies in Atlantic Canada, *Geophys. J.*, **39**, 385–406, 1974.

EDWARDS, R. N. and J. P. GREENHOUSE, Geomagnetic variations in the eastern United States: Evidence for a highly conducting lower crust?, *Science*, **188**, 726–728, 1975.

HUTTON, V. R. S., J. M. SIK, and D. I. GOUGH, Electrical conductivity and tectonics of Scotland, *Nature*, **266**, 617–620, 1977.

JACOBI, R. and Y. KRISTOFFERSEN, Geophysical and geological trends on the continental shelf off northeastern Newfoundland, *Can. J. Earth Sci.*, **13**, 1039–1051, 1976.

KOIZUMI, C. J., A. RYALL, and K. F. PRIESTLEY, Evidence for a high-velocity lithospheric plate under northern Nevada, *Bull. Seismol. Soc. Am.*, **63**, 2135–2144, 1973.

LAW, L. K. and R. P. RIDDIHOUGH, A geographical relation between geomagnetic variation anomalies and tectonics, *Can. J. Earth Sci.*, **8**, 1094–1106, 1971.

MCKERROW, W. S. and L. R. M. COCKS, The location of the Iapetus Ocean suture in Newfoundland, *Can. J. Earth Sci.*, **14**, 488–495, 1977.

PARKHOMENKO, E. I., *Electrical Properties of Rocks*, 314 pp., Plenum Press, New York, 1967.

RAST, N., M. J. KENNEDY, and R. F. BLACKWOOD, Comparison of some tectonostratigraphic zones in the Appalachians of Newfoundland and New Brunswick, *Can. J. Earth Sci.*, **13**, 868–875, 1976.

RODGERS, J., *The Tectonics of the Appalachians*, 397 pp., Wiley-Interscience, New York, 1970.

ROKITYANSKIY, I. I. and I. M. LOGVINOV, An electrical conductivity anomaly on the Kirovograd block of the Ukrainian Shield, *Phys. Solid Earth (Izv.)*, **6**, 405–409, 1972.

STRONG, D. F., W. L. DICKINSON, C. F. O'DRISCOLL, B. F. KEAN, and R. K. STEVENS, Geochemical evidence for an east-dipping Appalachian subduction zone in Newfoundland, *Nature*, **248**, 37–39, 1974.

WILLIAMS, H., The Appalachians in northeastern Newfoundland—A two sided symmetrical system, *Am. J. Sci.*, **262**, 1137–1158, 1964.

WILLIAMS, H. and R. K. STEVENS, The ancient continental margin of eastern North America, in *The Geology of Continental Margins*, edited by C. A. Burk and C. L. Drake, pp. 781–796, Springer-Verlag, New York, 1974.

WILLIAMS, H., M. J. KENNEDY, and E. R. W. NEALE, The northeastward termination of the Appalachian orogen, in *The Ocean Basins and Margins, II*, edited by A. E. M. Nairn and F. G. Stehli, pp. 79–123, Plenum Press, New York, 1974.

WRIGHT, J. A., Anisotropic apparent resistivities arising from non-homogeneous, two-dimensional structures, *Can. J. Earth Sci.*, **7**, 527–531, 1970.

WYLLIE, P. J., *Ultramafic and Related Rocks*, 464 pp., Wiley and Sons, New York, 1967.

Magnetovariational and Magnetotelluric Investigations in S. Scotland

V. R. S. HUTTON and A. G. JONES*

Department of Geophysics, University of Edinburgh, Edinburgh, U. K.

(Received November 22, 1977; Revised January 23, 1978)

A two-dimensional array of 20 Gough-Reitzel magnetometers was operated over S. Scotland in 1973 and in 1974–5 magnetotelluric and magnetovariational observations in the period range 10–10,000 s were made in the same region. In this paper, the analyses of the magnetic data from both studies are presented in the form of induction vectors and hypothetical event contours. They suggest that the lateral variations in electrical conductivity structure associated with the Eskdalemuir anomaly are more complex than suggested by earlier studies. A marked discontinuity in electrical structure is apparent in a narrow belt parallel to and south of the S. Uplands fault. This belt is associated with a major gravity anomaly and with steep gradients in the seismic profile at crustal depths. Another discontinuity is indicated near the Northumberland Basin. Representative examples of the magnetotelluric analysis and of one-dimensional Monte Carlo inversion of the M-T data are presented for the three regions separated by these discontinuities. They show that the conducting zone associated with the Eskdalemuir anomaly is at a depth greater than 24 km, while on either side of this region, there are good conductors within crustal depths.

1. Introduction

The tectonic history of S. Scotland is currently a subject of much interest. The existence of an ocean—the Iapetus—in this region during early Palaeozoic times is now generally accepted on geological grounds (DEWEY, 1974; PHILLIPS *et al.*, 1976) and there is convincing palaeomagnetic evidence (BRIDEN *et al.*, 1973) that the separation between the continental masses of Scotland and England was eliminated by the end of the Caledonian orogeny. The plate tectonic processes associated with the Iapetus closure are, however, still uncertain and many tectonic models have been postulated to explain the surface geology (MOSELEY, 1977).

Some knowledge about the deep structure in this region has resulted from recent geophysical studies. For example, the Lithospheric Seismic Profile in Britain (LISPB) has indicated that the Moho discontinuity changes in character from a well-defined transition under the northern part of the Midland Valley to an ill-defined one under the Southern Uplands (BAMFORD *et al.*, 1976, 1978). The seismic section published by

* Present address: Institut fur Geophysik der Westfalischen Wilhelms Universitat, D-4400 Munster/Westf., West Germany.

these authors also shows that the top of the 6.4 km s^{-1} refractor dips sharply from about 7 km to more than 14 km under the S. Uplands fault, where LAGIOS and HIPKIN (1979) have found a tectonically significant gravity anomaly.

Prior to the studies reported in this paper, electromagnetic induction techniques had been used in this area by several workers. JAIN and WILSON (1967) interpreted their magnetotelluric observations from Eskdalemuir using curve fitting techniques. This yielded a three-layer geo-electric section in which there was a middle conducting layer with its upper interface at a depth of about 10 km. Interpretation of similar observations across the Irish Sea from Stranraer also gave a conductive zone at lower crustal depths. Magnetovariational observations, by EDWARDS *et al.* (1971) and others, have been interpreted in terms of an anomalous Southern Uplands conductor at either lower crustal or upper mantle depths. BAILEY and EDWARDS (1976) have more recently used Bailey's hypothetical event technique to suggest that this anomaly may be caused by "unusually low resistivity in rocks presently *in the lower crust* which formed part of a 'Proto-Atlantic' oceanic crust in the late Palaeozoic".

A two-dimensional array of Gough-Reitzel magnetometers has now been operated over southern Scotland and magnetotelluric soundings have been made at 13 locations across and along the region of the anomaly, referred to by previous workers as the Eskdalemuir Anomaly.

2. Data Acquisition and Analysis

The locations, at which the Gough-Reitzel 3-component variometers were installed and operated simultaneously for approximately three months between 1973–4, are shown on the map of S. Scotland—Fig. 1. The magnetotelluric sites are indicated by crosses on the same map—in this case, single station observations were made, a typical recording

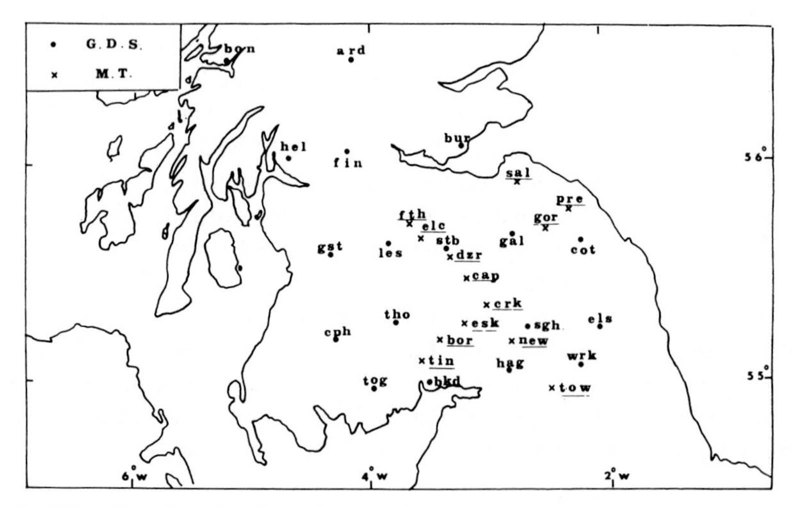

Fig. 1. Map of southern Scotland showing the location of: (a) the Gough-Reitzel magnetometer sites ●. (b) the magnetotelluric sites ×, names underlined.

interval at any one site being approximately 20 days for observations of periods in the 10 s to 24 hr range. Variations of the vertical component of the magnetic field, as well as the horizontal magnetic and electric field variations required for the magnetotelluric analysis, were recorded at these 13 M-T sites.

As the Gough-Reitzel magnetometers have a sensitivity of about 1 nT (GOUGH and REITZEL, 1967) and a digitisation interval of 20 s was used in this study, useful magnetovariational data was obtained from the array sites for periods $T>200$ s. Both fluxgate and JOLIVET (1966) magnetic sensors were employed in the magnetotelluric system which has been described by JONES (1977). A circuit diagram has been published by HUTTON (1976, Fig. 13).

Several techniques have been applied to the interpretation of data from these two investigations. In this paper emphasis is placed on the presentation and interpretation of the magnetovariational data using techniques which could be applied to both studies and by which a unified interpretation could be made of observations in the overlapping period range. The mapping of induction vectors and of hypothetical event contours suitably satisfy these objectives. A full account of the magnetotelluric study is published elsewhere (JONES and HUTTON, 1979 a, b). Examples of the results of the M-T data anal-

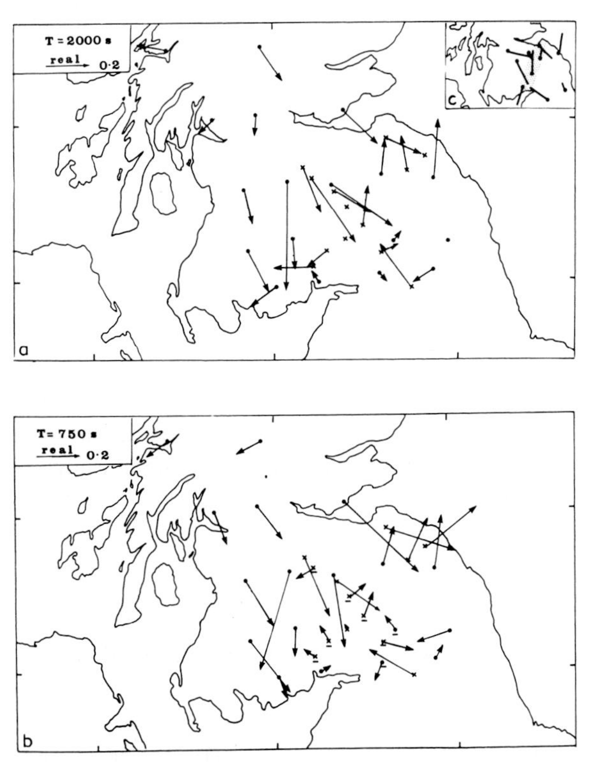

Fig. 2. The real 'Parkinson' vectors for (a) $T=2{,}000$ s and (b) $T=750$ s. The inserted Fig. 2 (c) has been redrawn from EDWARDS *et al.* (1971). It shows (a) their real 'Parkinson' vectors for $T=40$ min and (b) their proposed conductivity anomaly—dotted region.

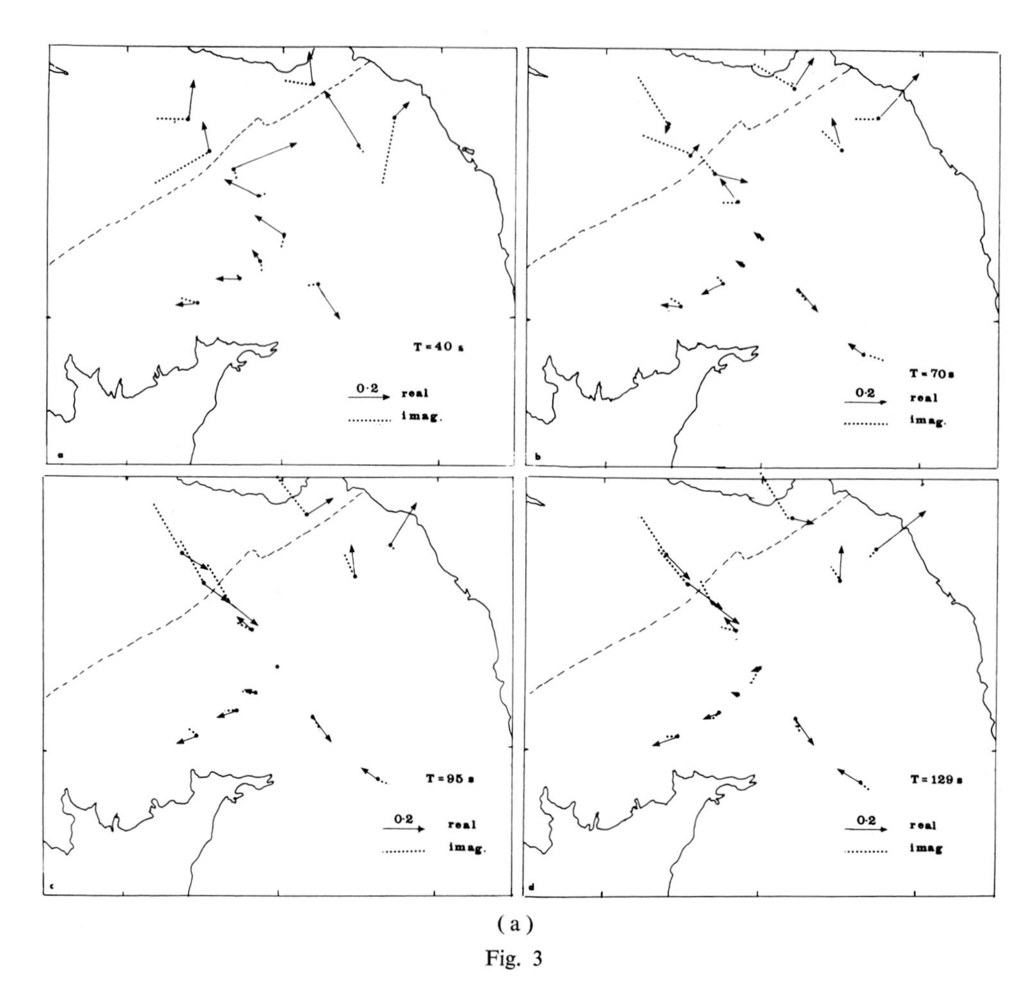

(a)
Fig. 3

ysis and of one-dimensional Monte Carlo inversion are presented here to illustrate the complementary nature of the two types of study and as an aid to the understanding of the magnetovariational results. A full account of the data analysis procedures is given by JONES (1977), who applied rigorous acceptance criteria to all the single station data from the M-T sites.

3. Magnetovariational Data

The real induction vectors from both studies are presented in Fig. 2(a) for $T=2{,}000$ s, and Fig. 2(b) for $T=750$ s. Following the Parkinson convention, their directions have been reversed so that they point towards current concentrations. The insert in Fig. 2(a)—Fig. 2(c)—has been redrawn from EDWARDS *et al.* (1971) and shows their real induction vectors for $T=40$ min and also a dotted region to indicate the strike of their proposed anomalous conducting layer. While, in general, the interpretation of our induction vectors for 2,000 s does not differ greatly from those of the earlier study, our map for

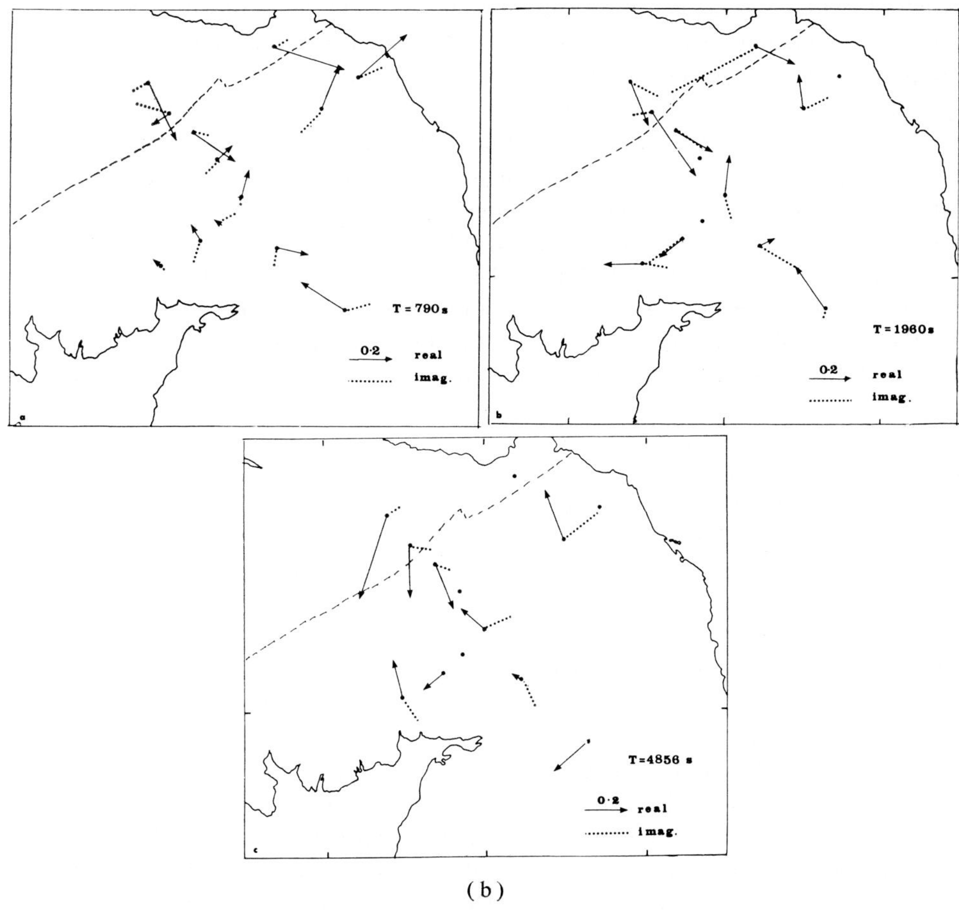

(b)

Fig. 3. Real and imaginary 'Parkinson' vectors for (a) T=40–129 s and (b) T=790–4,856 s. The location of the Southern Uplands Fault is denoted on each map by the dashed line.

the shorter period—Fig. 2(b)—includes a number of sites, denoted by underlined dots and crosses, at which the behaviour of the induction vectors is not compatible with a single conducting region as indicated in Fig. 2(c). Instead there is some suggestion at this period that there may be two regions of current concentration. Some insight into the complex structure responsible for the behaviour of these induction vectors can be obtained by following the change in pattern of the vectors as the period increases from the shortest period for which reliable data are available. The real induction vectors obtained at the M-T sites for periods in the range T=40 s to T=4,856 s are shown in Figs. 3(a) and (b). The location of the Southern Uplands Fault (S.U.F.) is also marked on each of the maps presented in these figures. At a period of 40 s, the vectors indicate current concentrations both to the north of the S.U.F. and to the south of Scotland. As the period increases, the vectors near the S.U.F. rotate until for periods of 95 s and 129 s, there are well-defined reversals in the directions of the vectors in (a) a narrow belt parallel to and south of the S.U.F. and (b) between the two most southern stations. The maps of vectors for periods of 790 s, 1,960 s, and 4,856 s also show changes in both the amplitude and direction of the real induction vectors at most sites. A current concentration near the S.U.F. appears to exist at all periods but that in the region of the Scottish border appears to be replaced at the longer periods by one further south, as indicated by the behaviour of the most southern vector. The existence of at least two anomalous regions is also suggested by the hypothetical event contours shown in Figs. 4(a)–(c). Using the technique first suggested by BAILEY *et al.* (1974), contours of equal vertical magnetic field amplitude have been drawn—for each figure—for a hypothetical uniform horizontal inducing field in a direction perpendicular to the strike of the major geological features of the region. In each map, large negative gradients in Z amplitude occur (a) near the S.U.F. and (b) near the Scottish-English border, and are compatible with lateral discontinuities in electrical conductivity structure in these two regions.

4. Magnetotelluric Data

Plots of apparent resistivity amplitude and phase as a function of period for 9 of the 13 sites could be classified into three distinct types, such as shown in the examples in Fig. 5. Data from the other 4 were rejected. Reference to Fig. 5 shows that stations FTH and SAL lie north of the S.U.F. and stations NEW, BOR, and PRE are in the Southern Uplands. Station TOW is in the Northumberland Basin. The errors associated with the estimates plotted in these figures are omitted in this presentation for the sake of clarity. It should also be noted that some of the average values plotted in Fig. 5, e.g. the 3 shortest period averages at SAL, are averages of 3 or less estimates. These values were not considered in the subsequent data inversion. For one station in each group, the results of a one-dimensional 3-layered Monte Carlo inversion (JONES and HUTTON, 1979 b) of the acceptable 'rotated major' apparent resistivity and phase data are shown schematically in Fig. 6. The method of rotation used in this study follows closely that suggested by REDDY and RANKIN (1974). The 'rotated major' direction can be regarded as equivalent to that of E-polarisation at stations on the conductive side of a vertical interface and of H-polarisation on the resistive side. The differences between the profiles presented in this figure indicate that there are major lateral variations in electrical conductivity structure in this region.

5. Conclusions

1) North of the Southern Uplands Fault there is conducting layer ($\sigma \sim 2\text{–}5 \times 10^{-2}$ S m^{-1}) extending from a depth of about 10 km to 45–65 km.

2) In the Southern Uplands, the whole of the crust is resistive and there is a good conductor *at upper mantle depths.*

3) A very good conducting layer $\sigma \sim 5 \times 10^{-1}$ S m^{-1} exists at TOW in the upper crust.

4) Lateral variations in electrical conductivity structure, as indicated by the magnetovariational data, are most marked in the narrow belts separating the three regions described in 1)–3) above. The possibility that this may be indicative of sharply dipping conductors connecting the three regions is now being examined by further mathematical

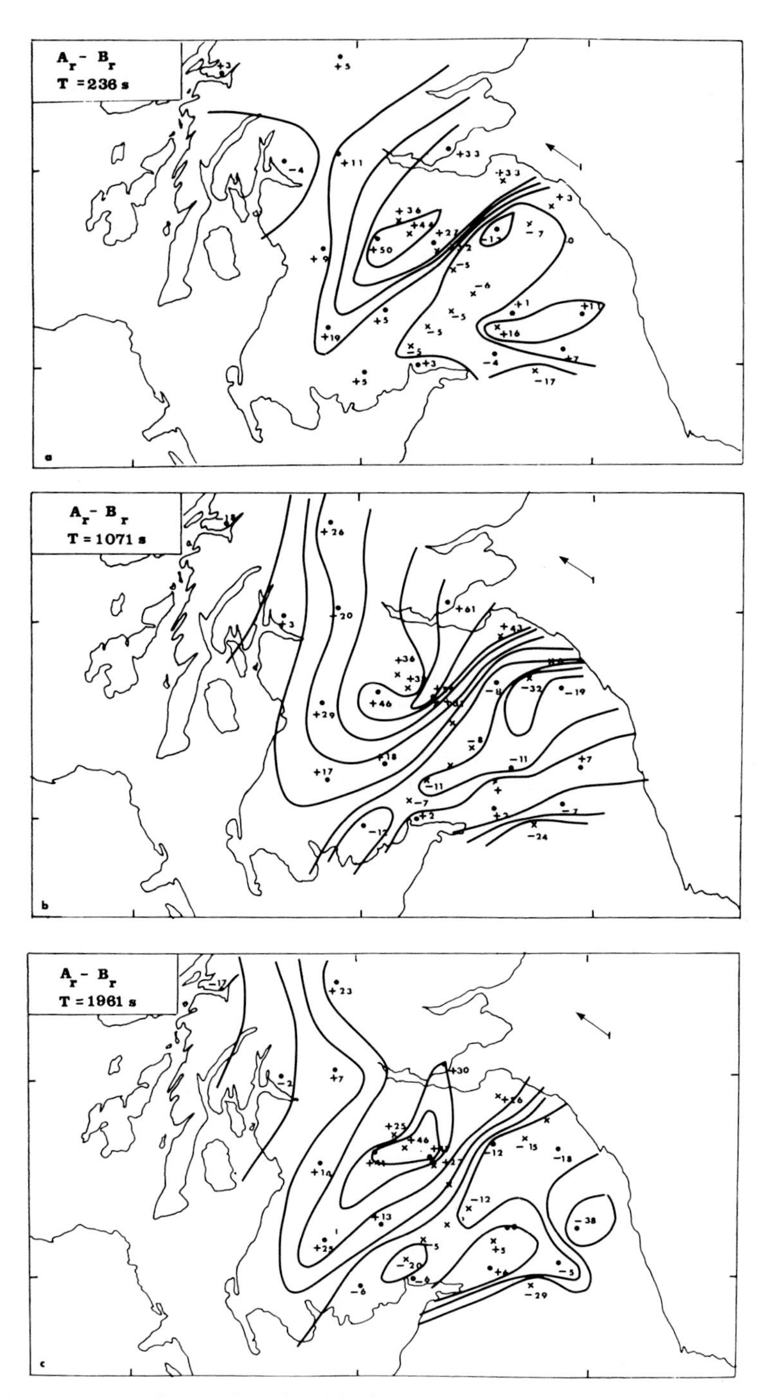

Fig. 4. Z contours for a unit regional horizontal inducing field directed as shown by the arrow drawn in the North Sea, i.e. approximately perpendicular to the strike of the major geological features of the area. (a) $T=236$ s, (b) $T=1{,}071$ s, and (c) $T=1{,}961$ s. The contour interval is 0.10 nT.

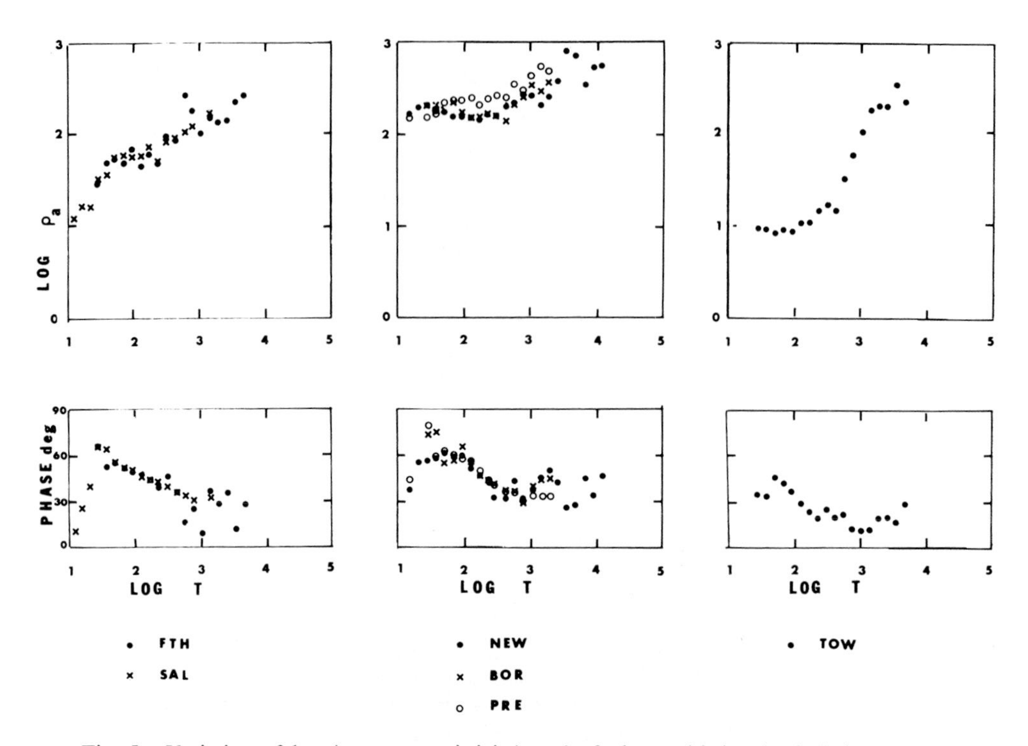

Fig. 5. Variation of log (apparent resistivity) and of phase with log (period) for several sites. FTH and SAL are north of the Southern Uplands Fault. NEW, BOR, and PRE are in the Southern Uplands and TOW is in the Northumberland Basin. ρ_a: Ωm, T: sec.

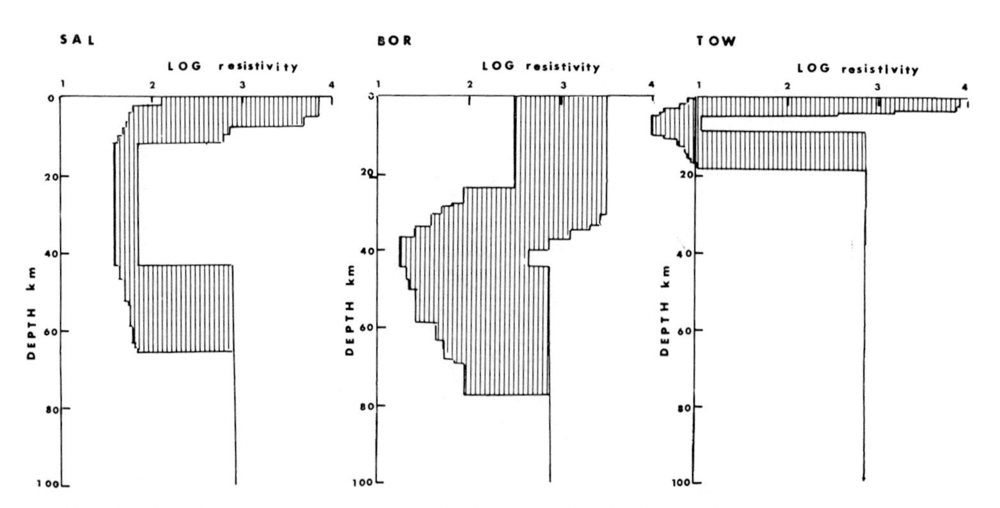

Fig. 6. A schematic representation of the resistivity-depth profiles obtained by one-dimensional Monte Carlo inversions of the apparent resistivity and phase data for SAL, BOR, and TOW. Note that the inversion of the BOR data demands a low resistivity middle layer (σ<90 Ωm). Resistivity: Ωm.

modelling of the data.

5) The marked lateral variation in electrical conductivity structure a few kilometers south of the Southern Uplands Fault occurs in a region where the top of the 6.4 km s^{-1} seismic refractor dips sharply and where there is a well-defined gravity low. The geophysical significance of these associations and also of the interpretation of the conductivity estimates which have been obtained will be discussed in a subsequent paper.

The authors acknowledge the support for this project of the Natural Environment Research Council, both in the form of research grants and of a studentship to one of us (A. G. J.). They are also indebted to Dr. D. I. Gough for the loan of his magnetometers, to Dr. J. Sik, Mr. S. Chita, and many others who participated in the fieldwork programme and to Dr. D. Rooney for assistance with the processing of the array data. The project could not have been undertaken without the willing cooperation of the many people who provided facilities for our observations.

REFERENCES

Bailey, R. C. and R. N. Edwards, The effect of source field polarisation on geomagnetic anomalies in the British Isles, *Geophys. J. R. Astr. Soc.*, **45**, 97–104, 1976.

Bailey, R. C., R. N. Edwards, G. D. Garland, R. Kurtz, and D. H. Pitcher, Electrical conductivity studies over a tectonically active area in Eastern Canada, *J. Geomag. Geoelectr.*, **26**, 125–146, 1974.

Bamford, D., K. Nunn, C. Prodehl, and B. Jacob, Crustal structure of Northern Britain, *Geophys. J. R. Astr. Soc.*, **54**, 43–60, 1978.

Bamford, D., S. Faber, B. Jacob, W. Kaminski, K. Nunn, C. Prodehl, K. Fuchs, R. King, and P. Willmore, A lithospheric seismic profile in Britain I. —Preliminary results, *Geophys. J. R. Astr. Soc.*, **44**, 145–160, 1976.

Briden, J. C., W. A. Morris, and J. D. A. Piper, Palaeomagnetic studies in the British Caledonides—IV. Regional and global implications, *Geophys. J. R. Astr. Soc.*, **34**, 107–134, 1973.

Dewey, J. F., The geology of the southern termination of the Caledonides, in *The North Atlantic: The Ocean Basins and Margins*, Vol. 2, Plenum Press, New York and London, 1974.

Edwards, R. N., L. K. Law, and A. White, Geomagnetic variations in the British Isles and their relation to electrical currents in the ocean and shallow seas, *Phil. Trans. Roy. Soc. London*, **270**, 289–323, 1971.

Gough, D. I. and J. S. Reitzel, A portable 3-component magnetic variometer, *J. Geomag. Geoelectr.*, **19**, 203–215, 1967.

Hutton, V. R. S., The electrical conductivity of the Earth and Planets, *Rep. Prog. Phys.*, **39**, 487–572, 1976.

Jain, S. and C. D. V. Wilson, Magnetotelluric investigations in the Irish Sea and Southern Scotland, *Geophys. J. R. Astr. Soc.*, **12**, 165–180, 1967.

Jolivet, J., Ph. D. Thesis, University of Paris, 1966.

Jones, A. G., Geomagnetic induction studies in S. Scotland, Ph. D. Thesis, University of Edinburgh, 1977.

Jones, A. G. and R. Hutton, A multi-station magnetotelluric study in southern Scotland–1. Fieldwork, data analysis and results, *Geophys. J.R. Astr. Soc.*, **56**, 329–349, 1979a.

Jones, A. G. and R. Hutton, A multi-station magnetotelluric study in southern Scotland—II. Monte-Carlo inversion of the data and its geophysical and tectonic imdlications, *Geophys. J. R. Astr. Soc.*, **56**, 351–368, 1979b.

Lagios, E. and R. G. Hipkin, The Tweeddale Granite—a newly discovered batholith in the Southern Uplands, *Nature*, **280**, 672–675, 1979.

Moseley, F., Caledonian plate tectonics and the place of the English Lake District, *Geol. Soc. Am. Bull.*, **88**, 764–768, 1977.

Phillips, W. E. A., C. J. Stillman, and J. Murphy, A Caledonian plate tectonic model, *J. Geol. Soc. Lond.*, **132**, 576–609, 1976.

Reddy, I. K. and D. Rankin Coherence function for magnetotelluric analysis, *Geophysics*, **39**, 312–320, 1974.

An Analogue Model Study of Ocean-Wave Induced Magnetic Field Variations Near a Coastline

T. MILES and H. W. DOSSO

Department of Physics, University of Victoria, Victoria, B. C., Canada

(Received October 7, 1977; Revised January 12, 1978)

A laboratory analogue model, described previously by MILES *et al.* (1977), was used to study the effect of a sloping ocean floor with a shelf, and a step with a shelf, on the magnetic variations induced by ocean waves moving in a static magnetic field. The model measurements indicate that for a shallow ocean, the amplitudes of the magnetic variations are attenuated as the wave travels over the wedge, reach a minimum at the shelf edge, then again increase over the shelf. For the case of the step with a shelf maximum attenuation occurs at the shelf edge followed by an enhancement over the shelf. For both cases, the depth of the fluid over the shelf is the important factor in determining the behaviour of the induced field.

1. Introduction

The induction of magnetic fields by ocean waves in an infinite depth ocean or in a uniform finite depth ocean has been studied theoretically by many authors, including YOUNG *et al.* (1920), CREWS and FUTTERMAN (1962), WEAVER (1965), GROSKAYA *et al.* (1972), LARSEN (1973), PREISENDORFER *et al.* (1974), KLEIN *et al.* (1975), and PODNEY (1975). Induction problems involving an ocean with a sloping or an irregular bottom, or an ocean with a coastline, have not as yet been solved analytically nor treated by numerical techniques. One method of studying these problems involves the aid of a suitably scaled laboratory analogue model. The present work deals with a laboratory analogue model study of fields induced by waves over a shelving sea floor and shelf, and over a vertical step and shelf.

2. Model Measurements and Discussion of Results

The analogue model scaling conditions used in the present work were developed and described earlier (MILES *et al.*, 1977) and will not be treated here. In the same work, the model construction, the experimental techniques, as well as the validity of the model results for a finite depth model ocean, were discussed in detail.

In the present work, the scaling factors for conductivity, linear dimension, and frequency are respectively $\sigma'/\sigma=4.33\times10^{-6}$, $d'/d=3.77\times10^{3}$, $f'/f=1.60\times10^{-2}$, where primed quantities refer to geophysical parameters and unprimed to model parameters. Using these scaling factors, 1 cm in the model corresponds to 38 m in the ocean, a wavelength of 9.5 cm in the mercury simulates a wavelength of 358 m in the ocean, and a frequency of 4 Hz in the model simulates a frequency of 0.064 Hz in the ocean. For the wavelength of $\lambda=9.5$ cm, the mercury depth of 2.8 cm was less than $1/3\ \lambda$. The steady primary field used in the model was 57 gauss ($5.7\times10^{-3}\ T$). Waves in the mercury were generated by dropping a small horizontal lucite rod into the mercury from a fixed height of 0.5 cm. This wave generator was situated 5 cm from one end of the mercury tank. The magnetometer probe, used to measure the induced field, was positioned so as to be sensitive to the horizontal induced field, as measured in a direction normal to the wave front in the mercury. The probe, mounted on a rail, could be positioned very accurately. The height of the probe above the mercury was 0.6 cm. This corresponded to a height of 23 meters above the surface of the ocean.

Measurements of the induced field were carried out point by point for traverses over

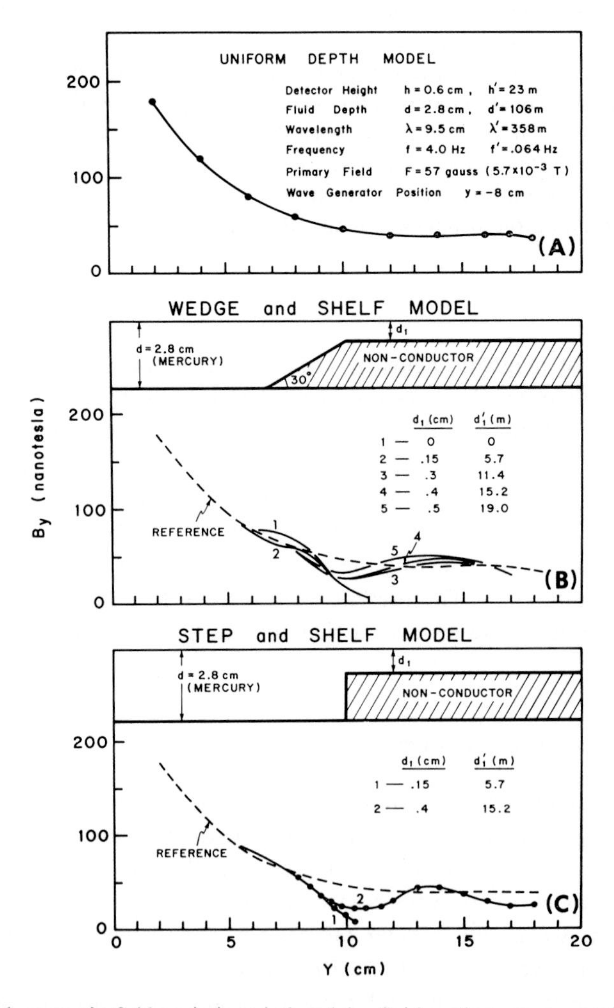

Fig. 1. Model magnetic field variations induced by fluid surface waves moving in a static magnetic field.

the model ocean for the cases of a uniform depth ocean, a sloping sea floor and shelf, and for a vertical step and shelf in the sea floor. The induced fields for these cases are shown in Fig. 1.

The induced field for a traverse over the uniform depth ocean is shown in Fig. 1(a). With the wave source located at $y=-8$ cm, it is seen that the induced field initially falls off rapidly with distance from the source then becomes essentially uniform in the neighbourhood of $y=10$ cm. The curve of Fig. 1(a) is used as a reference (dashed line) in Figs. 1(b) and (c).

To simulate a sloping ocean floor and a shelf near a coastline, a non-conducting (lucite) 30° wedge was positioned in the mercury as shown at the top of Fig. 1(b). With the uniform depth region held at a constant depth $d=2.8$ cm, measurements were made for various wedges with shelf depths d_1. The curves in this diagram show the magnetic field amplitudes for five different wedges and the curves labelled 1–5 correspond to shelf depths $d_1=0$, 5.7, 11.4, 15.2, 19.0 meters. It is to be noted for curve 1 that the amplitude of the induced field builds up over the ramp, becoming a maximum approximately half way up and then falls off sharply as the ocean depth decreases further. The enhancement begins just beyond the point towards the conductor where the wedge meets the ocean bottom. Somewhat similar behaviour, but without enhancement, is observed for the very shallow layer for curve 2, with the field falling off sharply again. For each of the other 3 curves corresponding to shelf depths of 11.4, 15.2, 19.0 meters, essentially no enhancement occurs over the ramp, but there is an attenuation with a field minimum directly over the edge of the shelf. The field then builds up over the shelf, attaining a value slightly greater than the reference field. The minimum value over the shelf edge decreases as the ocean depth over the shelf increases. As the depth d_1 is increased further, the fields should approach the reference values. The attenuation at the shelf edge, and the build up of the field beyond the shelf edge can be explained in part as due to the modification of the waves as they pass over the slope from the uniform depth ocean to the shallow ocean, and in part in terms of the changing induced current distribution with changing ocean depth. The increasing wave amplitude for a range of decreasing fluid depths would tend to enhance the field. Further, the interaction with the obstacles (wedge and step) may lead to interference patterns in the wave motion over the shelf, resulting in further induced field anomalies.

In Fig. 1(c) we see results for the case of a step in the ocean floor followed by a shelf for two depths $d_1=0.15$ cm and $d_1=0.4$ cm. For this model, a very small enhancement occurs as the wave approaches the vertical interface, the field falls off rapidly as the interface is approached for both curves. For the 0.4 cm depth, the broad attenuation has a minimum approximately over the interface, the minimum value being essentially the same as we saw for the case of the wedge for corresponding depth $d_1=0.4$ cm. The amplitude then increases to a maximum at approximately $3^1/_2$ cm beyond the interface corresponding to approximately $1/3\ \lambda$ beyond the interface, and then falls off to values below the reference value. Again, it is expected that as d_1 is increased further the fields should approach the reference curve values.

Measurements for further depths d_1 for the wedge and step cases will be carried out and presented at a later date. Other cases of a non-uniform sea floor, including sea mounts and dykes, too can readily be studied. In conclusion, the present results indicate that the wave motion and the induction mechanism is far from simple for the case of the non-uniform ocean floor, and that further studies are warranted.

REFERENCES

CREWS, A. and J. FUTTERMAN, Geomagnetic micropulsations due to the motion of ocean waves, *J. Geophys. Res.*, **67**, 299–306, 1962.

GROSKAYA, Ye. M., R. G. SKRYNNIKOV, and G. V. SOKOLOV, Magnetic field variations induced by the motion of sea waves in shallow water, *Geomagn. Aero.*, **12**, 131–134, 1972.

KLEIN, M., P. LOUVET, and P. MORAT, Measurement of electromagnetic effects generated by swell, *Phys. Earth Planet. Inter.*, **10**, 49–54, 1975.

LARSEN, J. C., An introduction to electromagnetic induction in the ocean, *Phys. Earth Planet. Inter.*, **7**, 389–398, 1973.

MILES, T., H. W. DOSSO, and T. P. NG, An analogue model for studying magnetic variations induced by ocean waves, *Phys. Earth Planet. Inter.*, **14**, 137–142, 1977.

PODNEY, W., Electromagnetic fields generated by ocean waves, *J. Geophys. Res.*, **80**, 2977–2990, 1975.

PREISENDORFER, R. W., J. C. LARSEN, and M. A. SKLARZ, Electromagnetic fields induced by plane-parallel internal and surface ocean waves, Report No. HIG-74-8, Hawaii Institute of Geophysics, University of Hawaii, 1974.

WEAVER, J. T., Magnetic variations associated with ocean waves and swell, *J. Geophys. Res.*, **70**, 1921–1929, 1965.

YOUNG, F. B., H. GERRARD, and W. JERONS, On electric distrurbances due to tides and waves, *Phil. Mag.*, **40**, 149–159, 1920.

Long Period Variations of the Geomagnetic Field and Inferences about the Deep Electric Conductivity of the Earth

A. M. IŞIKARA

Department of Geophysics, Faculty of Earth Sciences, University of Istanbul, Turkey

(Received February 22, 1978; Revised April 26, 1978)

The depth of penetration and the modified apparent resistivity are estimated by using the results of the annual and the solar cycle variation of the geomagnetic field, and then the deep conductivity is discussed.

1. Introduction

Electric conductivity (or its reciprocal resistivity) is one of the basic parameter of the earth which can be studied on a global basis. Its importance lies in the fact that below the crust the value is largely controlled by temperature. Estimates of the conductivity of the earth's mantle have been deduced from geomagnetic variations of external origin, from the secular variation and from core-mantle coupling.

In the frequency spectrum of geomagnetic variations of external origin, the annual and the solar cycle variation are the longest period variations. Since the longer period variations penetrate deeper, they have great potentials of probing the very deep conductivity structure within the earth's mantle.

Considerable works have been done to estimate the conductivity of upper mantle, but there are few works on the problem of the conductivity of lower mantle. Therefore, in the present note, the deep conductivity will be discussed by the combined results of these two low frequencies of geomagnetic spectrum, together with the results obtained for upper mantle.

2. Inferences about the Deep Conductivity

In the models of LAHIRI and PRICE (1939), the current induced by Sq and D_{st} do not penetrate appreciably beyond a depth of about 1,200 km, so that the knowledge of conductivity obtained from these variations is restricted to a certain depth. Estimates of conductivity at greater depths have been derived from investigations of the rate of change of the secular variation MCDONALD (1957). In such studies, the source field is unknown, all we have is the observed secular variation. The knowledge of conductivity at greater depths can also be obtained from the long term geomagnetic variations of external origin. In this case, it is possible to obtain the source and induced fields or in other words external and internal parts of the variation. The ratio of the parts of the magnetic field variation of internal and external origin is known as a measure of inductive response of the earth. Therefore, the first step is to determine this ratio for the annual and the solar cycle variation of the earth's magnetic field.

The ratio of internal to external parts of the annual variation was determined by MALIN and IṢIKARA (1976) using a more extensive and uniform data set than hitherto. They have shown that P_2^0 source field is the dominant term, but the other may also be present for the annual variation. Then, using the results for g_2^0 term as well as the weighted mean of (g_1^0, g_2^0, g_2^1) terms and assuming as g_2^0, the inductive scale length (the depth of penetration) is calculated (cf. Eq. 23, SCHMUCKER, 1970) for both terms after having the real and the imaginary parts of the ratio of internal to external parts of the annual variation.

The depth of penetration and the ratio of internal to external parts of the solar cycle variation are calculated by the transfer functions which the details were given in IṢIKARA (1977). A P_1^0 source field is assumed for the 11 years solar cycle variations.

The modified apparent resistivity obtained by $\rho^*=2\omega\mu_0\,(\mathrm{Im}(C))^2$ can be used as an estimator of the resistivity at the depth $z^*=\mathrm{Re}\,(C)$. Here ω is the frequency in cph, $\mu=4\pi\cdot10^{-7}$ volt·sec/Amp·m and C is the inductive scale length. It is found that ρ^* is

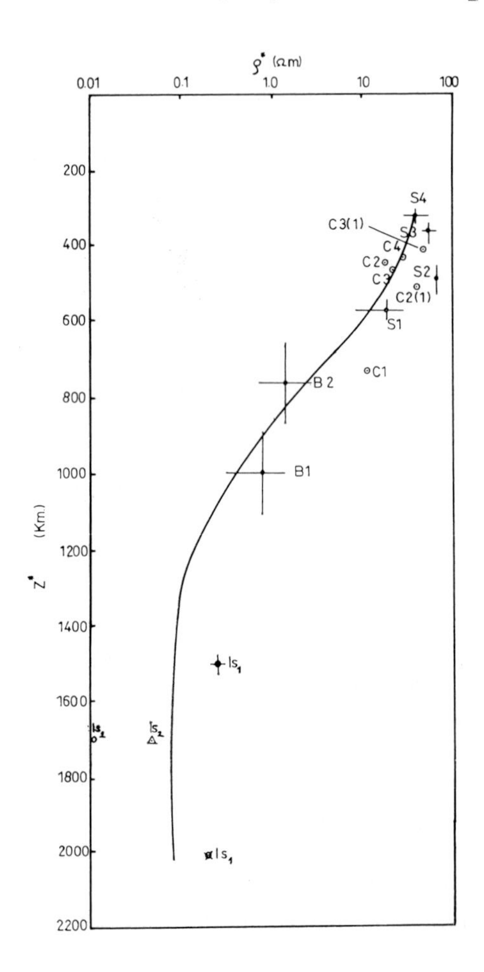

Fig. 1. Modified apparent resistivity against depth. Is_1 and Is_2 show the values obtained from solar cycle and annual variations, respectively.

0.19 Ωm at the depth of 1,500 km for the solar cycle variation. By using the equations 9 and 10 in ISIKARA (1977), the modified apparent resistivity is also obtained as 0.21 Ωm at the depth of 2,000 km. The ρ^* are 0.01 Ωm and 0.05 Ωm at the depth of 1,700 km for g_2^0 term and the weighted mean of (g_1^0, g_2^0, g_2^1) terms of the annual variation. BANKS and BULLARD (1966) used the annual variation to estimate a resistivity of about 0.5 Ωm at the depth of 1,275 km. As this result was obtained from 6 observatories data, it could be regarded as a doubtfull one. Although YUKUTAKE (1965) could not produce a reliable phase difference, he obtained a value of about 0.02 Ωm at a depth of 1,600 km, using magnetic variation data associated with the 11 years solar cycle. The depth of penetration closely agree with the present result, but the estimated resistivity here is greater with a factor of ten than Yukutake's.

The representation of the true resistivity distribution $\rho(z)$ by the modified apparent resistivity distribution $\rho^*(z^*)$ is usually adequate, if ρ decreases with increasing depth. The calculated values are shown in Fig. 1, together with the results obtained from *Sq* and D_{st} variations of the geomagnetic field (after HUTTON, 1976, Fig. 26). According to all these results, it can be reached to the conclusion that the apparent resistivity changes rapidly with increasing depth, but the gradient of the change in resistivity decreases considerably deeper than 1,200 km.

It has also been reached to the same conclusion from theoretical conductivity profiles, MCDONALD (1957) and YUKUTAKE (1959), that below 1,000 km depth the rate of change of conductivity with depth appears to be much less. Therefore, presented profile is in consistent with these theoretical profiles for the earth's mantle.

The more detailed study will be carried out by IŞIKARA and MALIN (1980) for the deep conductivity of the earth.

I am indebted to Prof. Dr. İ. Özdoğan for support and encouragement. I wish to thank Prof. Dr. U. Schmucker for his valuable advice and for his kind help.

REFERENCES

BANKS, R. J. and E. C. BULLARD, The annual and 27 day magnetic variations, *Earth Planet. Soc. Lett.* **1**, 118–120, 1966.

HUTTON, V. R. S., The electrical conductivity of the earth and planets, *Rep. Prog. Phys.*, **39**, 487–572, 1976.

IŞIKARA, A. M., Solar cycle controlled variation of the geomagnetic field, *Acta Geodaet, Geophys. et Montanist.*, **12**, 397–405, 1977.

IŞIKARA, A. M. and S. R. C. MALIN, Semi-annual variation of the geomagnetic field, 1980 (in preparation).

LAHIRI, B. N. and A. T. PRICE, Electromagnetic induction in nonuniform conductors, and the determination of the conductivity of the Earth from terrestrial magnetic variations, *Phil. Trans. Roy. Soc. London, Ser. A*, **237**, 509–540, 1939.

MALIN, S. R. C. and A. M. IŞIKARA, Annual variation of the geomagnetic field, *Geophys. J. R. Astr. Soc.*, **47**, 445–457, 1976.

MCDONALD, K. L., Penetration of the geomagnetic secular field through a mantle with variable conductivity, *J. Geophys. Res.*, **62**, 117–141, 1957.

SCHMUCKER, U., An introduction to induction anomalies, *J. Geomag. Geoelectr.*, **22**, 9–33, 1970.

YUKUTAKE, T., Attenuation of geomagnetic secular variation through the conducting mantle of the earth, *Bull. Earthq. Res. Inst. Tokyo Univ.*, **40**, 1–65, 1959.

YUKUTAKE, T., The Solar cycle contribution to the secular change in the geomagnetic field, *J. Geomag. Geoelectr.*, **17**, 287–309, 1965.

Inverse Magnetotelluric Problem for Sounding Curves from Czechoslovak Localities†

K. Pěč,* J. Pěčová,** and O. Praus**

*Geophysical Institute, Charles University, Prague 2, Czechoslovakia
**Geophysical Institute, Czechoslovak Academy of Sciences, Prague 4-Spořilov, Czechoslovakia

(Received September 25, 1979; Revised March 25, 1980)

In fitting geoelectrical models to sounding curves from certain localities in Czechoslovakia, we applied "the Hedgehog" method. In the first step it uses the Monte Carlo procedure to find a successful model. Then, it goes systematically through the values of both the electrical conductivities and the layer thicknesses varying within given limits in the neighbourhood of the successful model previously found.

The inversion was effected for different types of magnetotelluric sounding curves corresponding to specific geological structures. From a large family of models tested by the Hedgehog procedure the best fitting and the most frequently occurring models were chosen. The direct problem was solved for their parameters and the resulting theoretical sounding curves were compared with the experimental ones.

1. Introduction

In the preceding years the magnetotelluric soundings (MTS) were carried out at several temporary observatories along the DSS profile No. VI on the Czechoslovak territory (Fig. 1). The results of measurements and their interpretation were summarized in the paper by Pěčová *et al.* (1976).

This paper classifies the sounding curves obtained at individual MTS localities into three groups corresponding roughly to the principal geological units of the Czechoslovak territory. The first family of MTS curves, shown in Fig. 2, characterizes the geoelectrical conditions in basins with a thick layer of superficial sediments. These curves are typical of the Pannonian block of southern Slovakia and Hungary.

The second group of MTS curves (Fig. 3) was obtained in the Carpathian foredeep forming a transition zone between the Bohemian Massif and the Carpathians. This area has a rather complicated geological structure with dipping boundaries, overthrusts and faults, especially in its marginal parts. The third collection of MTS curves (Fig. 4) was obtained at sounding localities in the crystalline block of the Bohemian Massif with sediments of only a small thickness or usually with no sediments at all.

So far MTS curves have been interpreted approximately by estimating only the depths of layers with increased electrical conductivities from the straight-line portions of decreasing apparent resistivities. The object of our further work was to reinterpret

† Presented at the Fourth Workshop on EM Induction in the Earth and Moon, Murnau, September 1978.

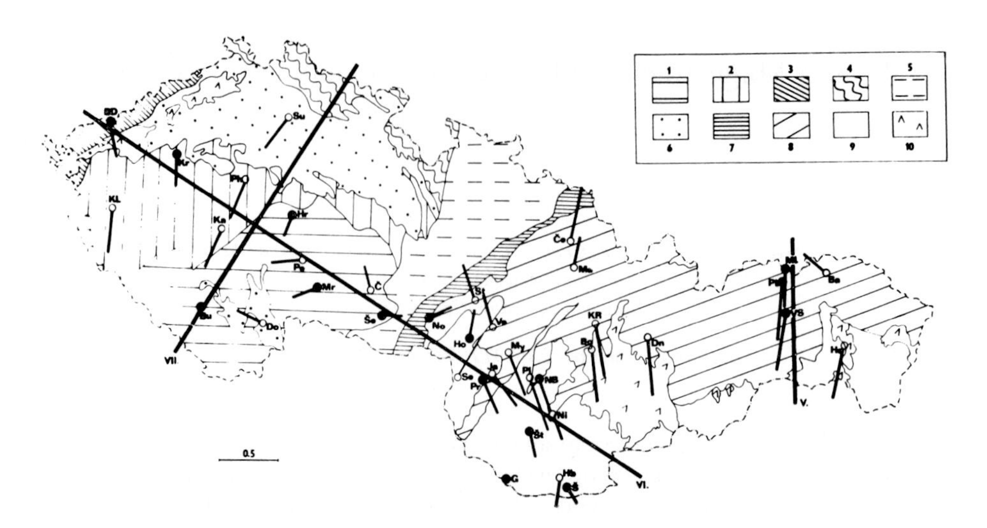

Fig. 1. Geological map of Czechoslovakia with DSS profile Nos. V, VI, and VII. Full circles indicate the Czechoslovak field stations, open circles specify observation sites of the GDR groups (H. Wiese and E. Ritter). Induction vectors are displayed as characteristics of electromagnetic induction in deep parts of different areas. The symbols for the stations the sounding curves of which were interpreted by the Hedgehog inversion: Š, Šrobárová; G, Gabčíkovo; Pr, Prievaly; Ho, Pánov (=Hodonín); Bu, Budkov. Geological symbols: Bohemian Massif; 1, Moldanubian block; 2, Assynthian block; 3, Saxo-Thüringian block; 4, West Sudeten block; 5, Moravian-Silesian zone; 6, Cretaceous and Tertiary platform deposits; 7, Alpine-Carpathian foredeep. Carpathian region; 8, pre-Neogene folding of the Carpathian unit; 9, Neogene basin of the Carpathian region; 10, Tertiary eruptive rocks.

these results by using the Hedgehog inversion method. Approaching the interpretation problem in this way, we are able to refine the preliminary interpretation and also to estimate the resistivities and thicknesses of individual layers.

2. Hedgehog Inversion of MTS Curves

The inversion problem of MTS curves was solved under the assumption that a model consisting of N horizontal, electrically homogeneous, isotropic layers was a good approximation of the internal geoelectrical structure. The bottom layer with subscript N was assumed to be a half-space of infinitely high electrical conductivity. Upon normalizing the resistivities and thicknesses of each individual layer to the corresponding values of the surface layer, we used the ratios $\bar{\rho}_i = \rho_i/\rho_1$ and $\bar{h}_i = h_i/h_1$ for dimensionless parameters of a specific model. The resistivity ρ_1 and thickness h_1 of the surface layer were estimated from independent geophysical, or geological information. Under these assumptions, our model was defined by $2(N-2)$ parameters $\bar{h}_2, \bar{h}_3, \cdots, \bar{h}_{N-1}, \bar{\rho}_2, \bar{\rho}_3, \cdots, \bar{\rho}_{N-1}$. The procedure for solving the direct problem was available so that we were able to calculate directly the theoretical master curve for any given set of parameters.

The problem is to choose the parameters in such a way that the solution of the direct problem may be as close as possible to the apparent resistivity curve obtained

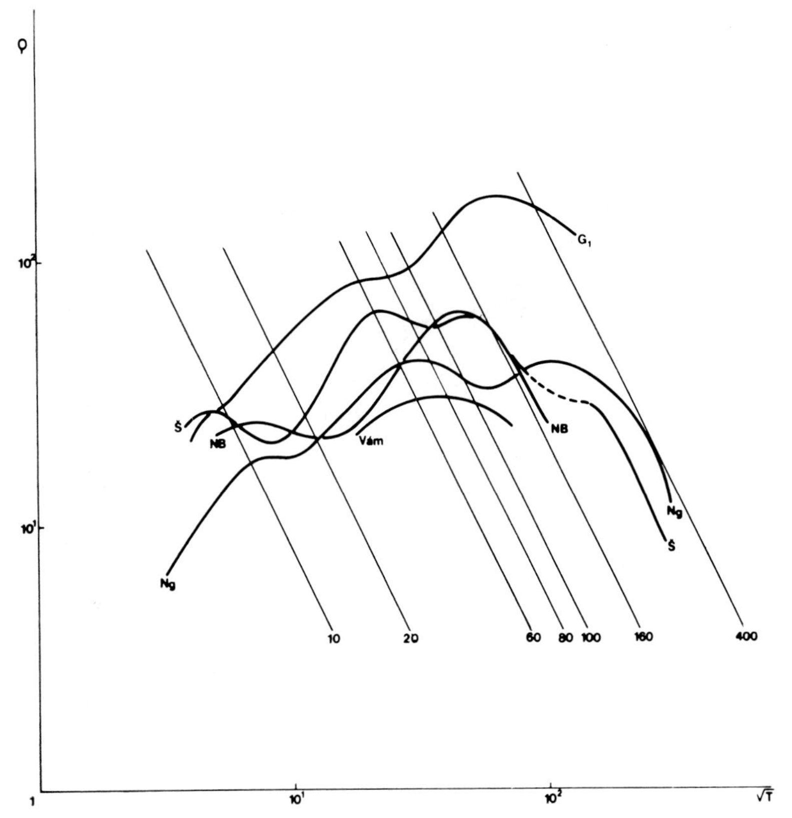

Fig. 2. MTS curves from the Pannonian basin. For the Hedgehog inversion the sounding curves from Šrobárová (Š) and from Gabčíkovo (G_1) were chosen.

from MTS measurements. To estimate the closeness of the fit, we defined a non-negative difference function $D(\bar{h}_i, \bar{\rho}_i)$. If it is zero, there is perfect coincidence of both the theoretical and experimental curves. It increases in value as the theoretical and experimental curves diverge. In our case, the mean square deviation was chosen for the difference function. Using it as a criterion, we may divide the computed sets of parameters into a class of successful models, which satisfy the condition $D(\bar{h}_i, \bar{\rho}_i) \leq A (A \geq 0)$ and into a class of rejected models which do not satisfy this condition.

In solving the inversion problem, first we had to specify roughly a region within which the parameters of a successful model can lie. For each parameter $p_i (p_1 = \bar{h}_2, \cdots, p_{N-2} = \bar{h}_{N-1}, p_{N-1} = \bar{\rho}_2, \cdots, p_{2(N-2)} = \bar{\rho}_{N-1})$ we thus define lower and upper $B_i \leq p_i \leq C_i$ $(i = 1, 2, \cdots, 2(N-2))$.

The constant A was estimated intuitively in the first step of the inversion procedure. In performing the computations by blocks consisting of roughly 500 successful models, we made the constant A smaller for each successive block to set the parameters of successful models closer to the local minimum of the difference function D. The initial values of the parameter bounds B_i, C_i were estimated by the trial and error method in

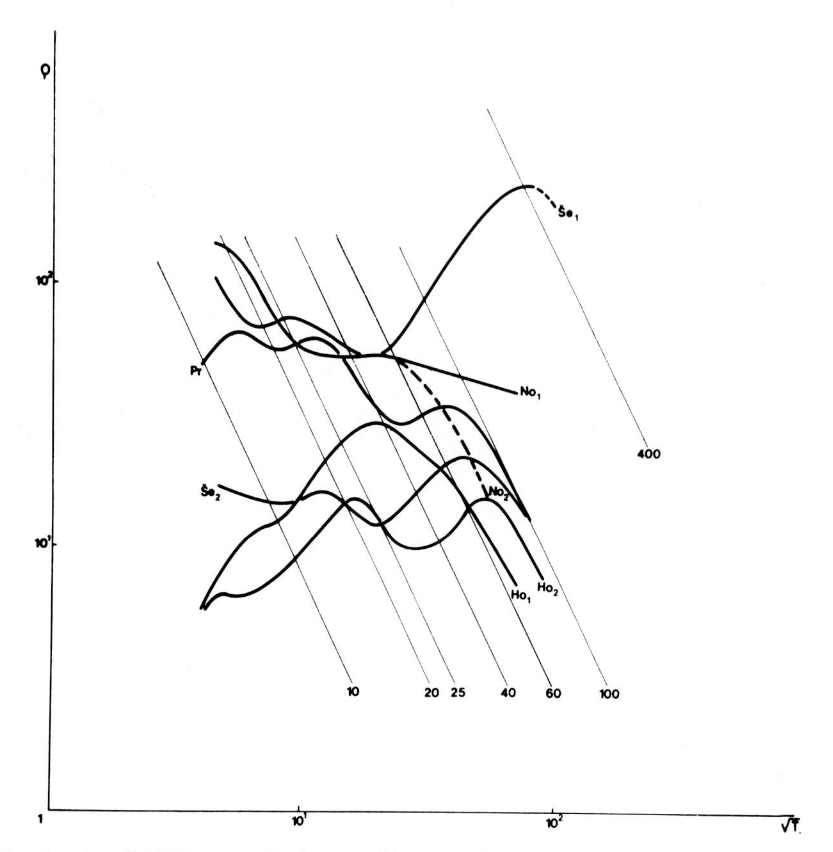

Fig. 3. Set of MTS curves in the transition zone between the Bohemian Massif and the Carpathians. The curves in the principal directions are displayed together with some apparent resistivity curves close to the principal directions. For the Hedgehog inversion the following sounding sites were chosen: Prievaly (Pr), Pánov (Ho).

advance and they were also changed during the computations to satisfy the conditions of the problem better.

For the inversion of MTS data, the Hedgehog method by KENNET (1972) was used. It was referred to first by VALJUS (1968) who applied it to the inversion of seismic data. For our application each interval (B_i, C_i) of individual parameters was divided into a grid of points sufficiently dense so that each parameter p_i $(B_i \leq p_i \leq C_i)$ might acquire only the values corresponding to the grid points, called the knots.

The Hedgehog inversion starts with the Monte Carlo method. The parameters p_i are randomly chosen within the intervals (B_i, C_i). The direct problem is computed for each set of randomly chosen parameters and the results are tested by the difference function criterion. If a set of parameters does not correspond to a successful model, the Monte Carlo procedure continues. As soon as a successful model is found, the control of the computations passes from the Monte Carlo subroutine to the Hedgehog method itself. Unlike the Monte Carlo method, which is blind in the sense that the search for the following model is independent of the preceding step, the Hedgehog method per-

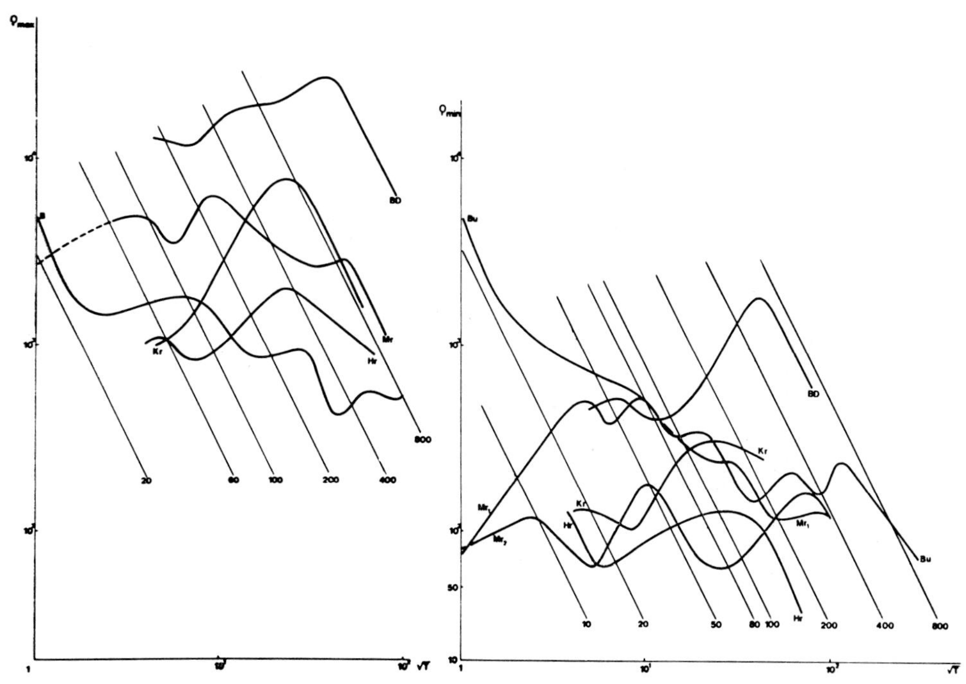

Fig. 4. Two sets of MTS curves ρ_{max} and ρ_{min} in principal directions for stations in the Bohemian Massif. The Hedgehog inversion was applied to sounding curve from station Budkov (Bu), ρ_{min}.

forms a systematic search for successful models in the neighbourhood of the first good knot. In other words, the Hedgehog method examines systematically the neighbouring knots and tests whether the corresponding solution is also successful.

3. Results of Inversion

Applying the Hedgehog inversion to MTS data, we obtained conductivity models for regions with a typical geoelectrical structure. The conductivity models displayed in Figs. 5 and 6 were obtained by inverting the MTS curves from localities of Šrobárová and Gabčíkovo. They both are typical of the geoelectrical conditions in the Pannonian basin covering southern Slovakia and a part of Hungary.

In Figs. 5 and 6, full lines define the most frequently occurring models in the last block of computations for the parameters close to the difference function minimum. Dashed lines characterize the best fitting models. Dot-and-dashed lines set the bounds for the depths and conductivities of individual layers. The width differs with the specific model and it depends not only on the constant A, but also on the scatter of experimental data and on the compatibility of actual geoelectrical conditions in the basement with theoretical assumptions (lateral homogeneity, horizontal layering etc.).

From the parameters obtained by inverting the MTS results of Šrobárová and

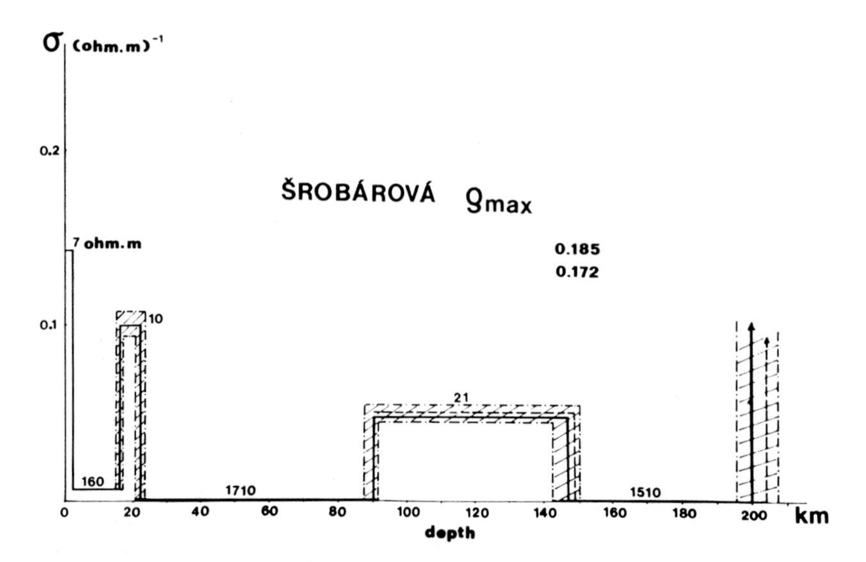

Fig. 5. Conductivity-depth model for the station of Šrobárová. Two figures below the station name are the sums of squared deviations for a given constant A, the upper one relating to the whole block of computations, the lower one corresponding to the best fitting model. Specific resistivities (in ohm·m) are shown for each layer.

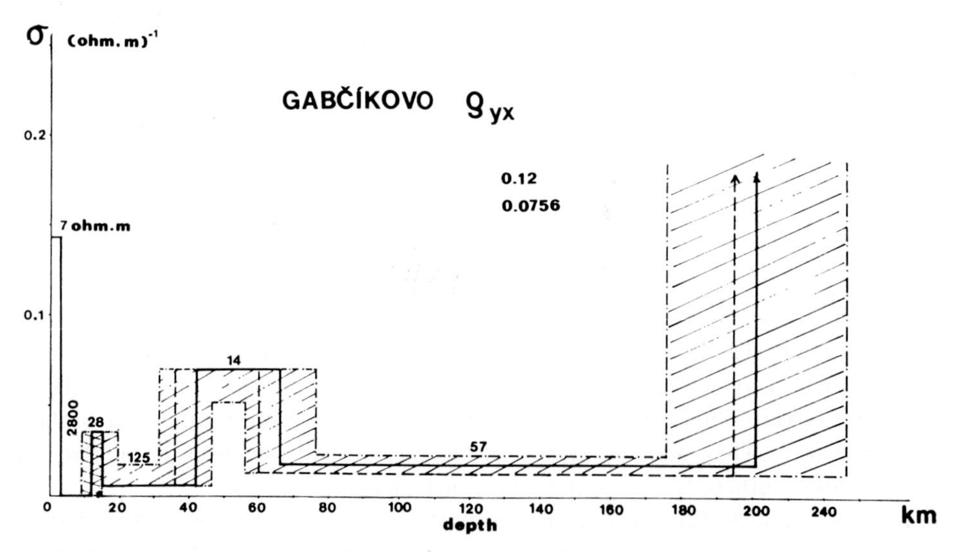

Fig. 6. Conductivity-depth model for the station of Gabčíkovo in the same form of presentation as in Fig. 5.

Gabčíkovo, the theoretical master curves were computed according to the direct problem formula. They are plotted in Figs. 7 and 8 together with the original experimental curve to be inverted. The best fitting and the most frequently occurring models are marked by crosses and open circles, respectively.

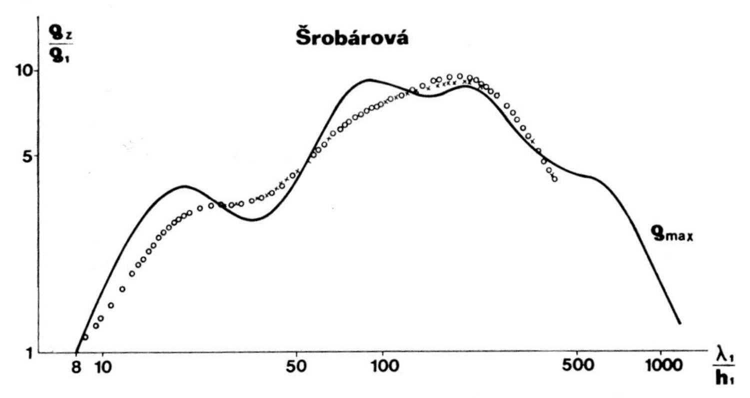

Fig. 7. Normalized experimental MTS curve $\rho_{\max}$ in the azimuth of maximum resistivity for the field station of Šrobárová (full line). The curves marked by open circles and crosses represent the theoretical curves for the most frequently occurring and the best fitting model, respectively. They were computed by the direct problem formula for the parameters obtained by the Hedgehog inversion (ρ_2, apparent resistivity; ρ_1, top layer).

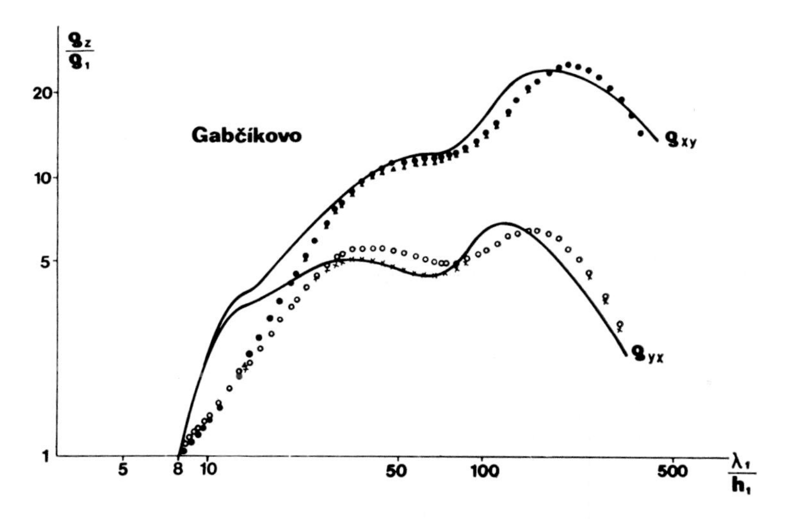

Fig. 8. Normalized experimental MTS curves ρ_{xy} and ρ_{yx} (in geographical NS and EW directions) for the field station of Gabčíkovo. Theoretical curves obtained by inversion are marked by full circles and triangles for ρ_{xy} and by open circles and crosses for ρ_{yx}, for the same types of models as in Fig. 7.

The model suggested for Šrobárová is characterized by three layers of increased electrical conductivity at the depths of 16, 90, and 200 km. There is good agreement concerning the depths, thicknesses, and resistivities with the model for Nagycenk observatory suggested by Ádám (1976). The sounding curves for Gabčíkovo station show an effect of strong anisotropy known also from several Hungarian sounding sites (Ádám *et al.*, 1968; Ádám and Verö, 1970).

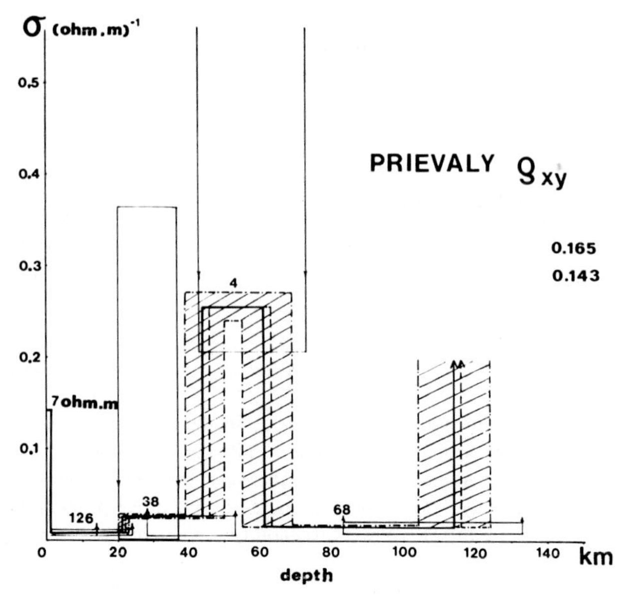

Fig. 9. Model of electrical conductivity-depth distribution for the field station of Prievaly displayed in the same way as in Fig. 5.

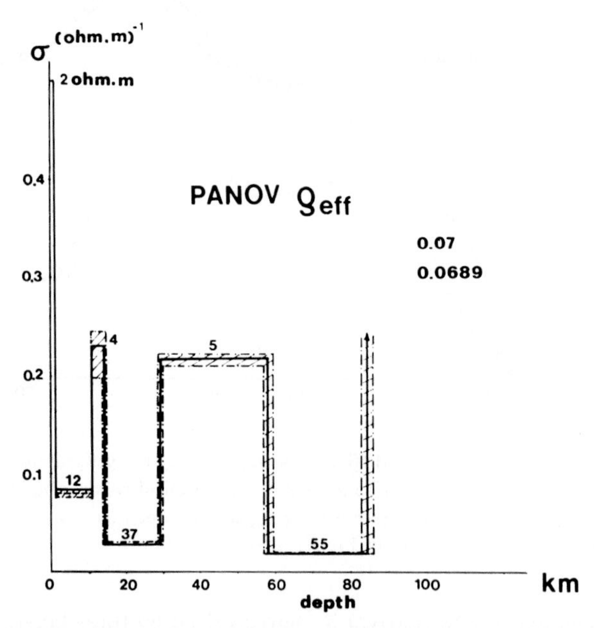

Fig. 10. Model of electrical conductivity-depth distribution for the field station of Pánov (=Hodonín) displayed in the same way as in Fig. 5. It corresponds to the sounding curve $\rho_{\text{eff}}=(\rho_{xy}\cdot\rho_{yz})^{1/2}$.

In the next step we applied the Hedgehog inversion to MTS results from stations at Prievaly and Pánov, both situated in the Carpathian foredeep. The conductivity

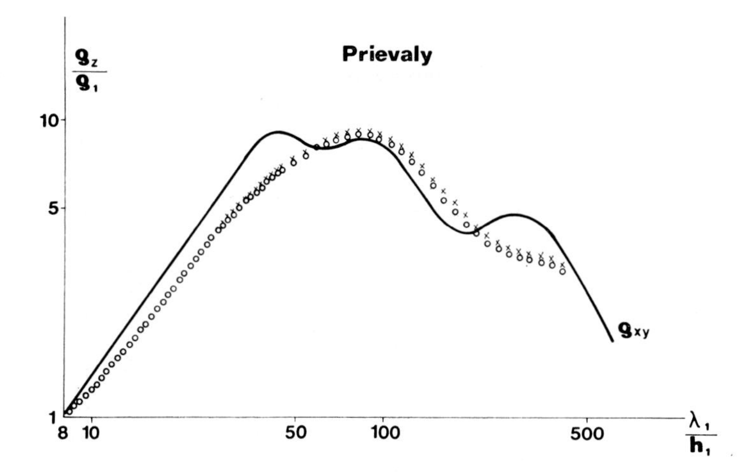

Fig. 11. Normalized experimental MTS curve ρ_{xy} for the field station of Prievaly together with the curves obtained by the Hedgehog inversion. They are shown in the same way as in Fig. 7.

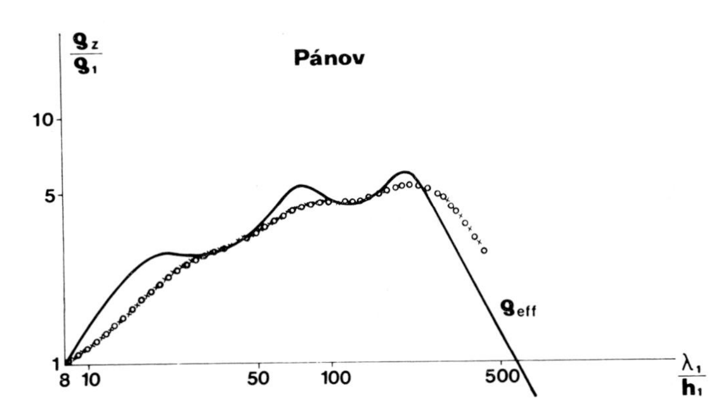

Fig. 12. Normalized experimental MTS curve ρ_{eff} for the field station of Pánov (=Hodonín) together with the curves corresponding to the results of the Hedgehog inversion. They are presented in the same way as in Fig. 7.

models (Figs. 9 and 10) and the theoretical master curves (Figs. 11 and 12) are presented similarly as in the previous case. In the geoelectrical cross-section of the Carpathian foredeep we find, according to the results from Prievaly and Pánov, very pronounced conductive layers at the depths of 42 and 30 km. They both have practically the same electrical conductivity of about 0.2 and 0.25 S/m (=(ohm·m)$^{-1}$). A very conductive asthenospheric layer is suggested at the depths of 115 and 85 km, fitting well the neighbouring estimates.

In the final step we applied the inversion procedure to MTS curve from the geomagnetic observatory of Budkov situated in the Bohemian Massif. A sounding curve of quite a different type was obtained here. It corresponds to a high resistive geoelectrical structure of an old crystalline shield, with large penetration depths of the inducing

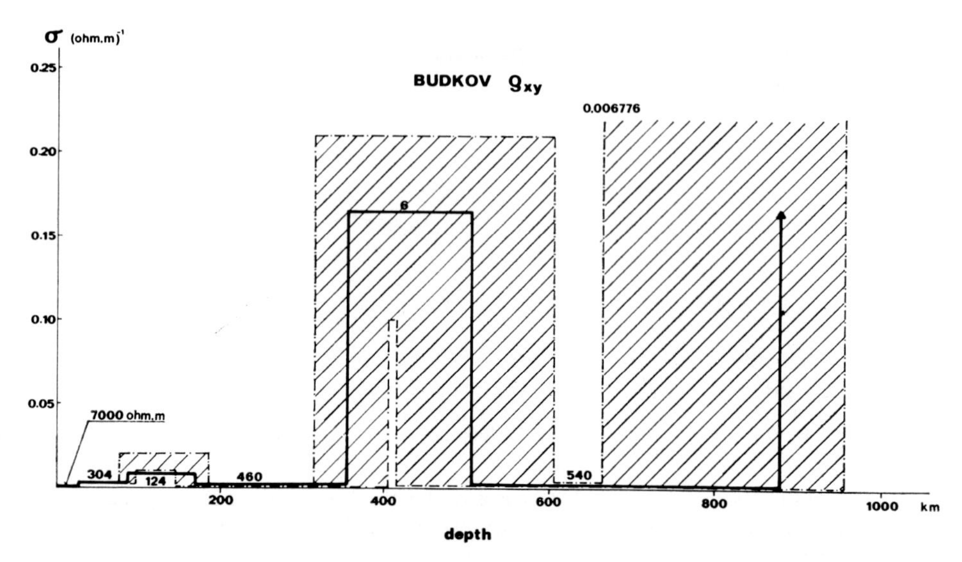

Fig. 13. Model of electrical conductivity-depth distribution for the electromagnetic observatory of Budkov (crystalline shield area) displayed in a similar way as in Fig. 5 (only for the most frequently occurring model).

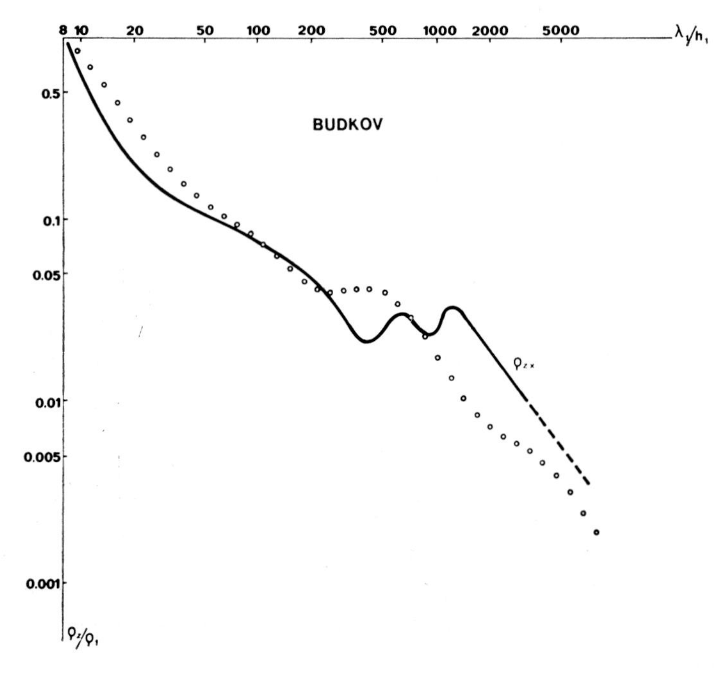

Fig. 14. Normalized experimental curve $\rho_{xy}=\rho_{zx}$ for the electromagnetic observatory of Budkov together with the theoretical curve corresponding to the results of the Hedgehog inversion (the most frequently occurring model).

field.

The inversion results presented in Figs. 13 and 14 suggest the existence of an asthenospheric conductive layer at depths of about 90–100 km. The conductivity, however, seems to be substantially lower than the corresponding estimate for the Pannonian block (Šrobárová, 0.048 S/m; Budkov, 0.008 S/m). This seems to be compatible with the results of surface heat flow measurements, suggesting much lower temperatures at the depth of the Bohemian Massif than at the corresponding depths of the Pannonian block (Čermák *et al.*, 1976).

Deeper layers of increased electrical conductivity suggested by the inversion results of Budkov observatory at about 400 and 800 km seem to correspond well with the results obtained at several other observatories, estimated from both the magnetotelluric and global geomagnetic data.

4. Conclusion

The sounding curves obtained from mangetotelluric measurements at field observatories situated along the Czechoslovak part of the DSS profile No. VI were interpreted under the assumption that a model consisting of horizontal layers was a realistic approximation. The electrical resistivities and layer thicknesses were determined by the Hedgehog inversion for curves characterizing geoelectrical conditions in regions with a specific internal structure (the Pannonian basin, the Carpathian foredeep, the Bohemian Massif).

Layers with increased electrical conductivities were obtained at crustal and subcrustal depths, slightly varying between individual regions. Only an asthenospheric layer of increased electrical conductivity seems to appear in the relatively narrow interval of depths between 85 and 115 km, in all typical regions in Czechoslovakia.

REFERENCES

Ádám, A., Results of deep electromagnetic investigations, in *KAPG Geophysical Monograph, Geoelectric and Geothermal Studies*, edited by A. Ádám, pp. 547–560, Akadémiai Kiadó, Budapest, 1976.

Ádám, A. and J. Verö, Das elektrische Modell des oberen Erdmantels im Karpatenbecken, *Acta Geod., Geophys. et Montanist. Acad. Sci. Hung.*, **T. 5**, 5–20, 1970.

Ádám, A., T. Miletits-Cz., and J. Verö, Magnetotelluric deep-soundings in Hungary, *Acta Geod., Geophys. et Montanist. Acad. Sci. Hung.*, **T. 3**, 129–147, 1968.

Čermák, V., J. Pěčová, and O. Praus, Heat flow, crustal temperatures and geoelectrical cross-section in Czechoslovakia, in *KAPG Geophysical Monograph, Geoelectric and Geothermal Studies*, edited by A. Ádám, pp. 538–542, Akadémiai Kiadó, Budapest, 1976.

Kennet, B. L. N., The Cambridge Hedgehog package for geophysical inversion, Univ. of Cambridge, 1972.

Pěčová, J., V. Petr, and O. Praus, Depth distribution of the electric conductivity in Czechoslovakia from electromagnetic studies, in *KAPG Geophysical Monograph, Geoelectric and Geothermal Studies*, edited by A. Ádám, pp. 517–537, Akadémiai Kiadó, Budapest, 1976.

Valjus, V. P., Determination of seismic cross-sections from sets of observations, *Computational Seismology*, No. 4, Nauka, Moskva, 3–11, 1968 (in Russian).

Remarks on Spatial Distribution of Long Period Variations in the Geomagnetic Field over European Area[†]

J. PĚČOVÁ,* K. PĚČ,** and O. PRAUS*

*Geophysical Institute, Czechoslovak Academy of Sciences, Prague 4-Spořilov, Czechoslovakia
**Geophysical Institute, Charles University, Prague 2, Czechoslovakia

(Received September 25, 1979; Revised March 10, 1980)

Geomagnetic observatory data were analysed to obtain the spectral amplitudes and phases of the 27-day variation and its harmonics over a network of mostly European observatories, but extended to global scale by including some suitably distributed geomagnetic stations on other continents. The spectral amplitudes, estimated by different methods, were subjected to spherical harmonic analysis by expanding the geomagnetic potential function into a series of spherical harmonics up to the 2nd order and by calculating the corresponding coefficients. The contours of the H and Z components were then calculated to analyse their latitude-longitude dependence over European area. The moduli of electromagnetic response functions estimated from global data for individual harmonics were compared with the results obtained previously for European data only. The distribution of deviations of (Z/H) ratios defined by differences between the actual data and the theoretical fits were investigated. Using the theoretical basis of electromagnetic induction in stratified conductors, we estimated the depths of a perfect substitute conductor at those depths from characteristics of the 27-, 13-, and 9-day variations.

1. Introduction

In our previous studies (PĚČOVÁ *et al.*, 1977a, b) we dealt with spatial distribution of long period geomagnetic variations over the European area based on an analysis of the geomagnetic data of European stations only. Recently, the data have been supplemented by the results obtained from some suitably distributed geomagnetic observatories on other continents. In this contribution we present new results and compare them with the previous ones. We also estimate the electrical resistivities at the depths corresponding to geomagnetic variations with periods of 27, 13, and 9 days and include these results in the resistivity-depth profiles suggested by SCHMUCKER (1970a).

2. Data Analysis

The geomagnetic observatories of both hemispheres, the published data of which were used in the present analysis are listed in Table 1. As in our previous studies, the

[†] Presented at the Fourth Workshop on EM Induction in the Earth and Moon, Murnau, September 1978.

Table 1. List of observatories and their geomagnetic co-ordinates.

Observatory	Θ	Λ	Observatory	Θ	Λ
Lovö	31°54′	105°48′	Toledo	46°06′	74°42′
Nurmijärvi	32 06	112 36	Ebro	46 06	79 42
Stony Hurst	33 06	82 42	Odessa	46 18	111 06
Valentia	33 24	73 30	L'Aquila	47 06	92 54
Leningrad	33 48	117 18	Surlari	47 06	106 06
Rude Skov	34 06	98 30	Panagyurishte	49 12	103 24
Wingst	35 24	94 06	Almeria	49 24	75 18
Hartland	35 24	79 00	Victoria	35 48	293 00
Witteveen	35 54	91 12	Memambetsu	56 00	208 24
Niemegk	37 48	96 36	Teoloyucan	60 24	327 00
Dourbes	38 00	87 42	Kakioka	64 00	206 00
Minsk	38 30	110 24	M'Bour	68 42	55 00
Swider	39 24	104 36	Kanoya	69 30	198 06
Chambon la F.	39 30	84 24	Paramaribo	73 00	14 18
Belsk	39 36	104 00	Tatuoca	80 24	20 48
Průhonice	40 06	97 18	Moca	84 18	78 36
Budkov	40 54	96 24	Bangui	85 12	88 30
Fürstenflb.	41 12	93 18	Hollandia	102 30	210 18
Lvov	42 00	105 54	Apia	106 06	260 12
Wien	42 06	98 12	Tsumeb	108 12	82 48
Kiev	42 24	112 12	Pilar	110 12	4 36
Nagycenk	42 48	98 18	Tananarive	113 42	112 30
Hurbanovo	42 48	99 48	Hermanus	123 18	80 30
Tihany	43 42	99 06	Amberley	137 42	252 30
Logrono	43 54	77 12	Port-aux-Fr.	147 18	128 00

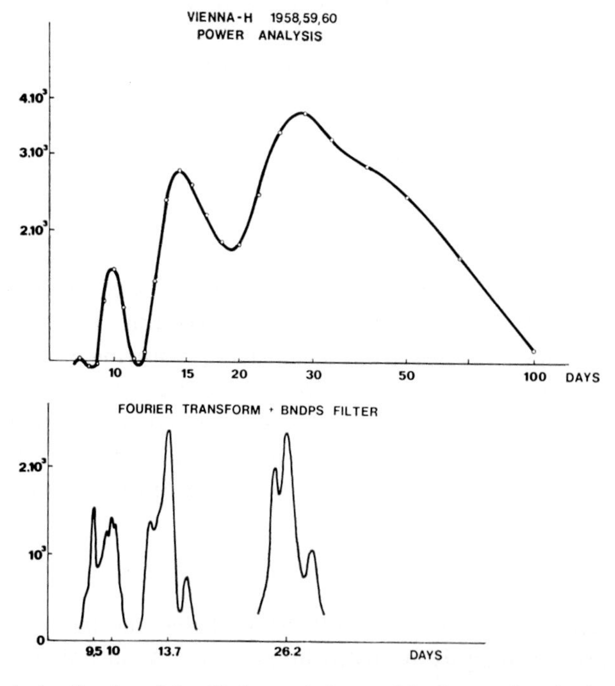

Fig. 1. Spectral estimates of the 27-day variation and its harmonics obtained by applying the power spectral analysis and the Fourier transform with band-pass filtering.

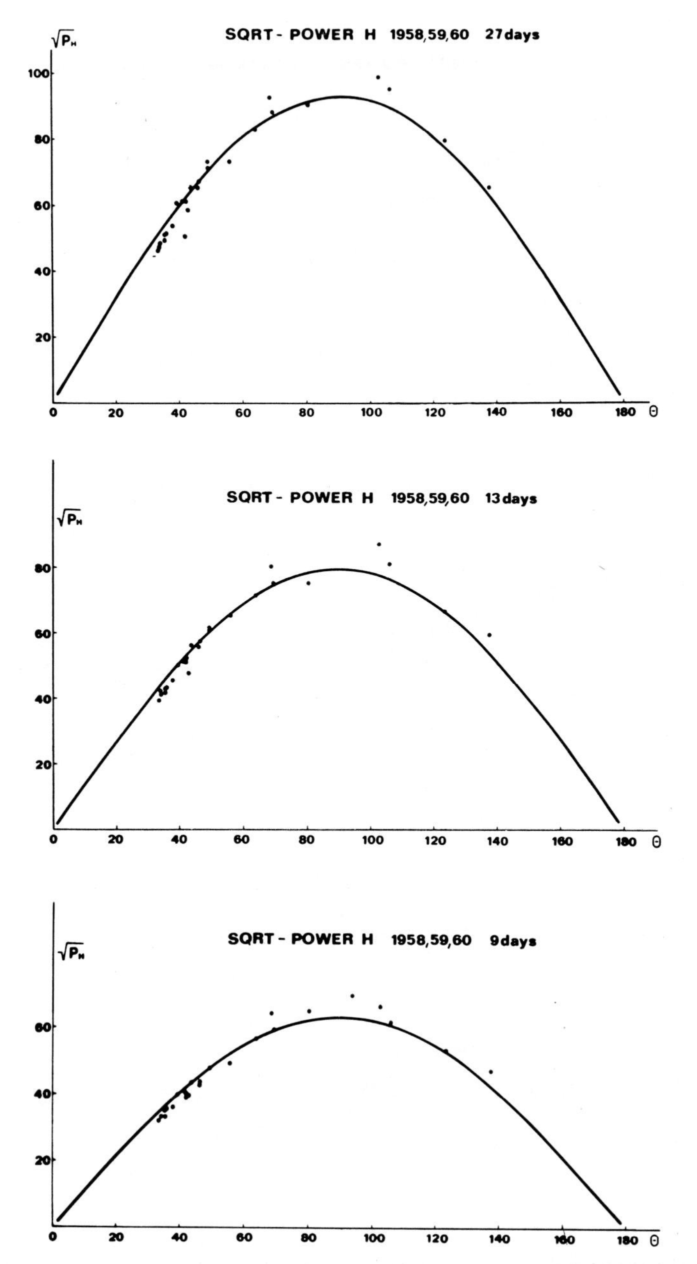

Fig. 2. The results of the least-squares fitting the 27-, 13-, and 9-day spectral estimates of the H component by function $A_{H,1}(dP_1(\cos\Theta)/d\Theta)=A_{H,1}\sin\Theta$. The unit of POWER-H is $(nT)^2/(c\cdot day^{-1})$.

daily means of the H, D, and Z geomagnetic components of three-year intervals of diverse solar activity were reanalysed by applying both the Fourier transform without or with a special band-pass filter and the power spectral analysis with a high-pass filter. In

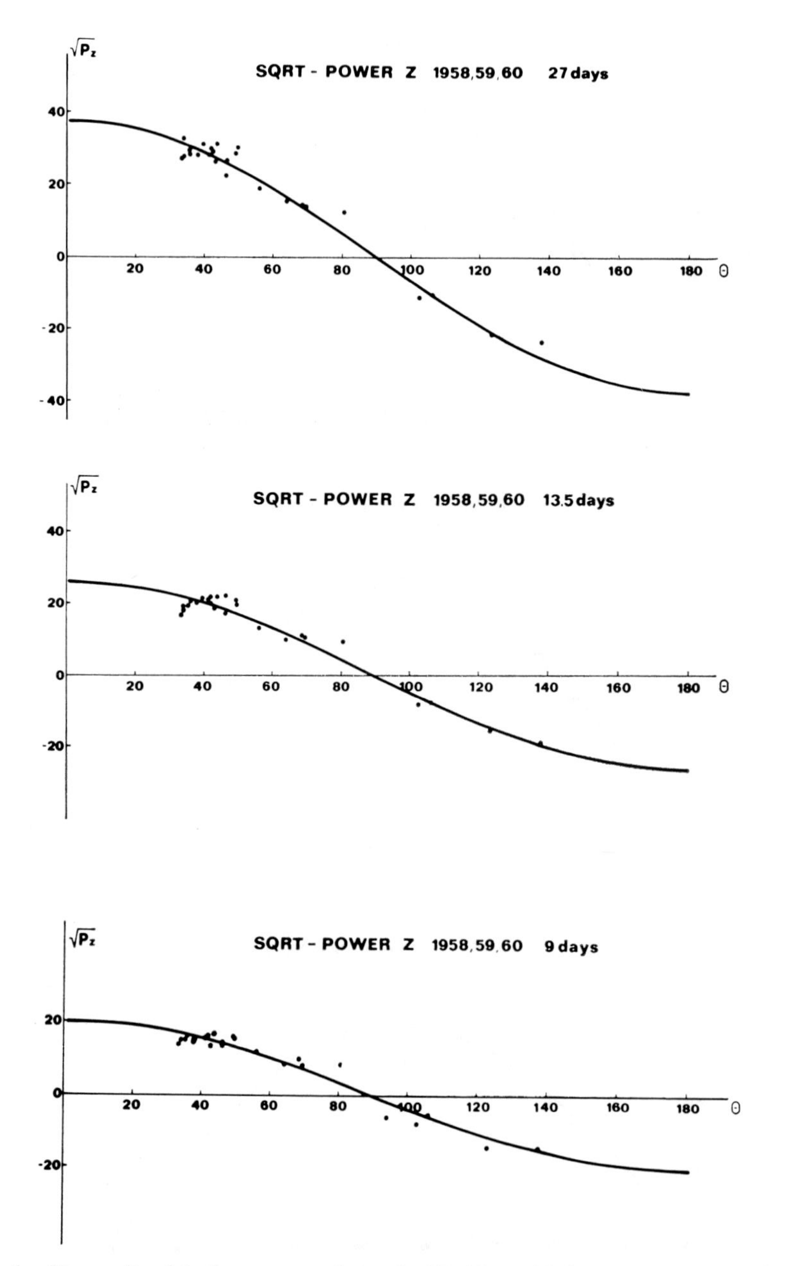

Fig. 3. The results of the least-squares fitting the 27-, 13-, and 9-day power spectral estimates of the Z component by function $A_{Z,1}P_1(\cos\Theta)=A_{Z,1}\cos\Theta$. The unit of POWER-Z is $(\mathrm{nT})^2/(\mathrm{c}\cdot\mathrm{day}^{-1})$.

Fig. 1 we present an example of spectral estimates obtained by both methods of analysing the data of the H component at the Vienna observatory during the interval 1958, 59, 60. The maximum values of these spectral estimates were subjected to further analysis.

Table 2. Estimates of the response function $|W_1|$.

Interval	FT			FT+BNDP			SQRT POWER		
	A_Z/A_H	FIT $A_Z A_H \pm \varepsilon_{FIT}$	$\varphi_Z-\varphi_H$	A_Z/A_H	FIT $A_Z/A_H \pm \varepsilon_{FIT}$	$\varphi_Z-\varphi_H$	A_Z/A_H	FIT $A_Z/A_H \pm \varepsilon_{FIT}$	$\varphi_Z-\varphi_H$
1958, 59, 60	0.393	0.400±0.068	172°	0.385	0.394±0.063	174°	0.405	0.416±0.039	164°
1963, 64, 65	0.308	0.295 0.104	186	0.306	0.294 0.103	168	0.340	0.329 0.085	166
1967, 68, 69	0.251	0.243 0.053	145	0.283	0.274 0.074	147	0.387	0.376 0.064	154
27–33 DAYS Average: $A_Z/A_H=0.338$, $\varphi_Z-\varphi_H=164°$, $Q_1=0.283$.									
1963, 64, 65	0.344	0.332±0.098	172	0.334	0.327±0.079	168			
25 DAYS Average: $A_Z/A_H=0.334$, $\varphi_Z-\varphi_H=170°$, $Q_1=0.285$.									
1958, 59, 60	0.294	0.293±0.052	151	0.304	0.305±0.051	153	0.328	0.333±0.036	161
1963, 64, 65	0.343	0.346 0.080	143	0.360	0.362 0.093	223	0.353	0.340 0.094	159
1967, 68, 69	0.268	0.253 0.060	163	0.286	0.275 0.059	144	—	— —	—
13–14 DAYS Average: $A_Z/A_H=0.318$, $\varphi_Z-\varphi_H=162°$, $Q_1=0.294$.									
1958, 59, 60	0.295	0.274±0.092	170	0.332	0.352±0.046	172	0.325	0.330± —	166
1963, 64, 65	0.244	0.236 0.080	162	0.419	— —	—	0.308	0.303 0.070	164
1967, 68, 69	0.262	0.254 0.066	228	0.307	0.301 0.064	—	0.311	0.302 0.035	164
8–10 DAYS Average: $A_Z/A_H=0.304$, $\varphi_Z-\varphi_H=175°$, $Q_1=0.302$.									

In the first approximation the spatial distribution was assumed to be purely zonal. It was obtained by the least-squares fitting the H and Z spectral estimates to functions $\sin\Theta$ and $\cos\Theta$, respectively, where Θ was the geomagnetic co-latitude. The results of the fitting the 27-, 13-, and 9-day power spectral peaks for the three-year interval of maximum solar activity are shown in Fig. 2 for the H component and in Fig. 3 for the Z component. The scatter of individual estimates over the best fitting curve is moderate and it remains practically the same for all other sets corresponding to different methods of analysis and intervals of diverse solar activity.

The amplitudes $A_{H,1}$ and $A_{Z,1}$ obtained from the best fits by functions $A_H \sin\Theta$ and $A_Z \cos\Theta$ were used to estimate the modulus of the ratio $|W_1|=|A_{Z,1}/A_{H,1}|$ which is of basic importance for calculating the response function $Q_1(f)$ according to the simplified formula by BANKS (1969)

$$Q_1(f)=\frac{V_1^{(i)}(f,R,\Theta)}{V_1^{(e)}(f,R,\Theta)}=\frac{1-[A_{Z,1}(f)/A_{H,1}(f)]}{2+[A_{Z,1}(f)/A_{H,1}(f)]}\,.$$

Function $Q_1(f)$ represents the ratio of the internal to the external parts of the geomagnetic variation potential. The ratios $(A_{Z,1}/A_{H,1})$, corresponding standard deviations ε for both the Fourier transforms and the power spectral estimates are summarized in Table 2 together with the average phase difference $\varphi_Z-\varphi_H$ between the variation components and with the mean values of the response function Q_1 for the principal periods. The scatter of values is caused by errors in spectral amplitude estimates determined by different approaches. It is also due to different levels of solar activity and to the shifts of spectral peaks in the Z component to periods slightly different from these of the H components. Since we have no reason to prefer special cases, we deal with the averaged values.

Comparing the recent results based on the global data with the previous ones obtained from European observations only (PĚČOVÁ *et al.*, 1977a, b), we can conclude that the both average estimates of the ratio $(A_{Z,1}/A_{H,1})$ differ very little. Thus, for internal geoelectrical structure, the European continent does not seem to be much different from the whole Earth.

3. Trend Analysis

For a detailed study of the spatial distribution different sets of H and Z spectral estimates were subjected to trend analysis performed by considering spherical harmonic terms of order $n=1, 2$ including the tesseral terms. Compared with a previous analysis, the fitting surface for $n=1$ differs by a tesseral term of the 1st order. Good agreement was found for the Z component, however, considerable differences found for the H component can be explained by the significant magnitude of tesseral terms.

The standard deviations were calculated for the trend surfaces of the 1st and 2nd orders. They are expressed in percentage of the mean input amplitude and compiled in Table 3. They show that the trend surface of the 2nd order fits the input amplitude distribution of the H component much better (about 10%) than does the 1st order surface (20%). In the Z component, the difference between the two fits is not so large, but also

Table 3. Standard deviations.

Case		H		Z	
		1st order	2nd order	1st order	2nd order
58–27	FT	21.0%	5.1%	24.8%	22.2%
58–27	FT+BNDP	21.0	—	23.4	24.1
58–27	$\sqrt{\text{POWER}}$	18.5	2.4	14.7	11.8
58–13	FT	17.8	5.5	22.4	15.4
58–13	FT+BNDP	18.0	6.6	22.2	19.8
58–13	$\sqrt{\text{POWER}}$	18.1	3.1	19.6	11.7
58–09	FT	16.4	4.8	36.3	25.5
58–09	FT+BNDP	23.3	6.3	20.4	19.0
58–09	$\sqrt{\text{POWER}}$	18.9	2.8	21.4	12.6
63–27	FT	18.1	18.3	39.5	31.7
63–27	FT+BNDP	18.6	10.6	37.5	22.0
63–27	$\sqrt{\text{POWER}}$	14.5	—	29.6	20.6
63–13	FT	18.2	8.6	26.2	21.2
63–13	FT+BNDP	18.9	7.6	28.8	21.4
63–13	$\sqrt{\text{POWER}}$	13.1	3.5	33.5	20.2

the 2nd order fit seems to be slightly better.

In order to demonstrate the trend surfaces by contour maps, we chose representative cases with small standard deviations (they are underlined in Table 3). Four contour maps in Fig. 4 show the trend surfaces of both orders for both Z and H components of the 27-day variation. Similarly, the contour map of 13- and 9-day variations in Fig. 5 characterizes the trend surfaces of the 2nd order fits.

All contours mapping the distribution of the horizontal component seem to support our previous results in suggesting that the dependence of the H variation field is mostly on the geomagnetic latitude.

The second order contours of the Z component, specifically those for the 13- and 9-day periods, clearly indicate the dependence of the vertical magnetic field on both geomagnetic co-ordinates. The latter result is also in good agreement with our previous conclusion (PĚČOVÁ *et al.*, 1977b). There are, however, some differences between both recent and former analyses. Having performed the trend analysis for the set of only European stations, we found that the 1st order fit was sufficient to account for the geomagnetic longitude dependence of the Z variation field. After extending the data to global dimensions and after performing the analysis, we find that the 1st order trend surface now characterizes the main features of the field distribution and only the 2nd order trend surface can account for regional European peculiarities.

Taking into account the new results, we checked our previous conclusions drawn from the distribution of deviations defined for the (Z/H) ratio over the European area (PĚČOVÁ *et al.*, 1977b). From the differences between the actually observed and the best fitting $Z/H=(A_{Z,1}/A_{H,1}) \cot \Theta$ ratios, the deviations were summed up and averaged for each European observatory at a specific co-latitude Θ. Applying a non-linear interpolation, we produced a contour map of the deviation pattern for the 27-day variation over the European region (Fig. 6). The general distribution of negative and positive

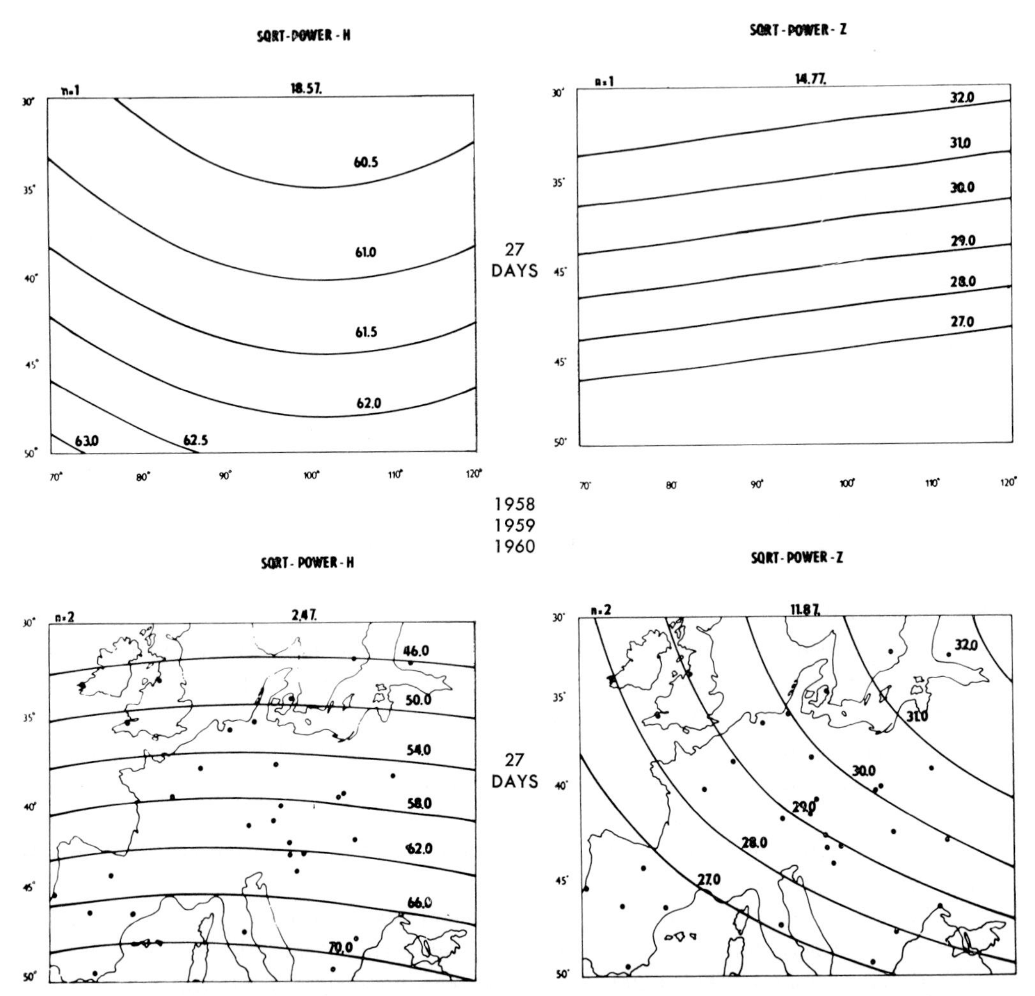

Fig. 4. Contour maps of both the H and Z power spectral estimates for the 27-day harmonic fitted to trend surfaces of the 1st and 2nd orders. The units of POWER-H and POWER-Z are $(\mathrm{nT})^2/(\mathrm{c \cdot day^{-1}})$.

deviations appears to be essentially similar to the previous one (PĚČOVÁ *et al.*, 1977b). Though the contours are preliminary, we can follow the distribution more closely. The European sector is divided into two parts, the northern with negative deviations and the southern with positive deviations. The zero-line traverses the northern part of Central Europe in the east-west direction. The area of the Baltic shield, Denmark, Northern Germany, the Netherlands, and the British Isles is characterized by negative deviations. Two singular points (isolated points with opposite sign), one in each part, are marked with circles.

The zero-deviation line may suggest a deep-seated zone with a laterally non-homogeneous distribution of the electrical conductivity in the upper mantle. No obvious correlation with parameters provided by other geophysical methods than geomagnetism

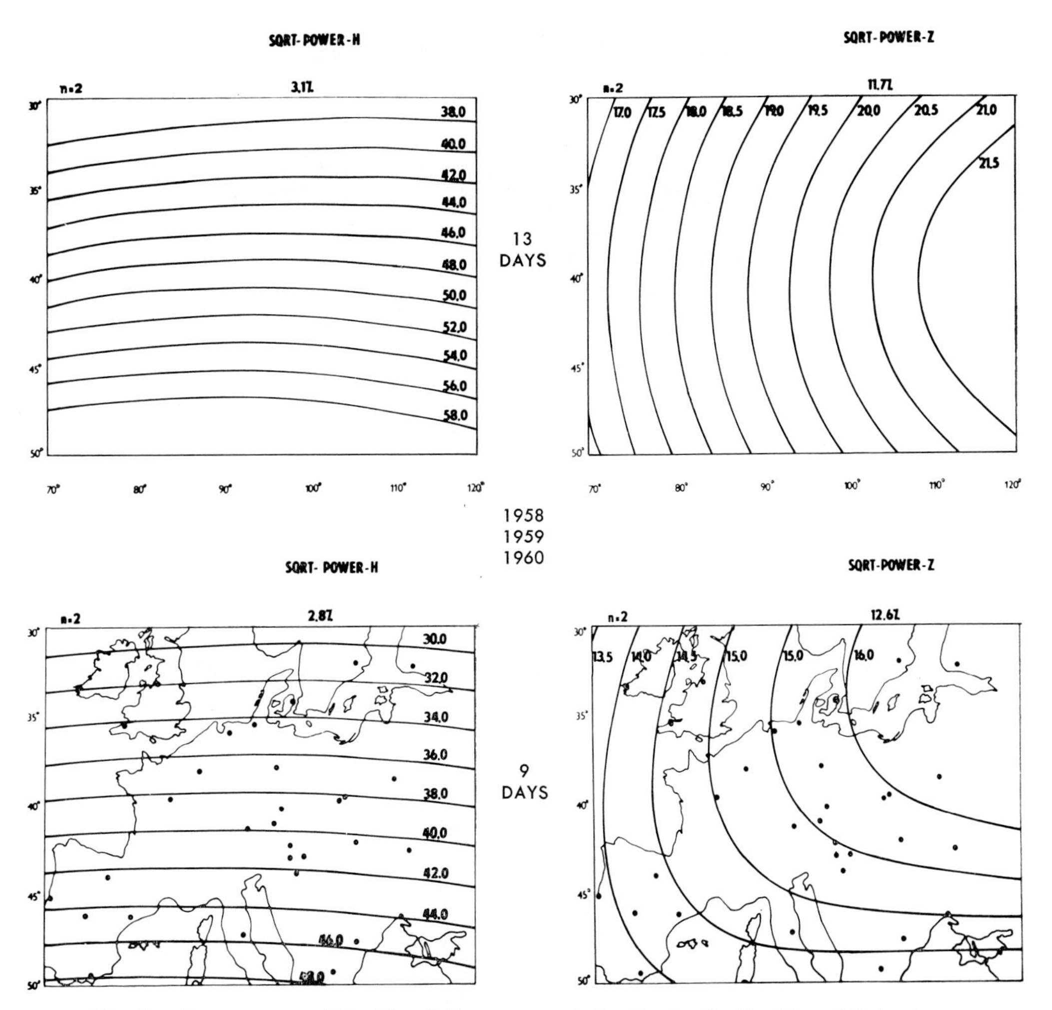

Fig. 5. Contour map of the H and Z power spectral estimates for the 13- and 9-day harmonics fitted by the 2nd order trend surfaces. The units of POWER-H and POWER-Z are $(\mathrm{nT})^2/(\mathrm{c}\cdot\mathrm{day}^{-1})$.

has been found so far, which corroborates this indication. It seems natural, therefore, to compare this result with the information deduced from other geomagnetic variations of long periods and of global dimensions.

The existence of the so-called "European anomaly in the geomagnetic variation field" is suggested by FAJNBERG *et al.* (1975) on the basis of the results obtained by analysing the S_q and D_{st} variations. In Fig. 7, reproduced from the paper by FAJNBERG *et al.* (1975), the radial component of an anomalous internal field induced by the 1st harmonic of the D_{st} variation is displayed in the form of a contour map. The anomalous zone is marked by a belt (5 to 10 degrees wide), within which the radial anomalous field changes from positive values in the south and south-west to negative values in the north and north-east. We can observe a rough spatial and directional similarity of this belt with

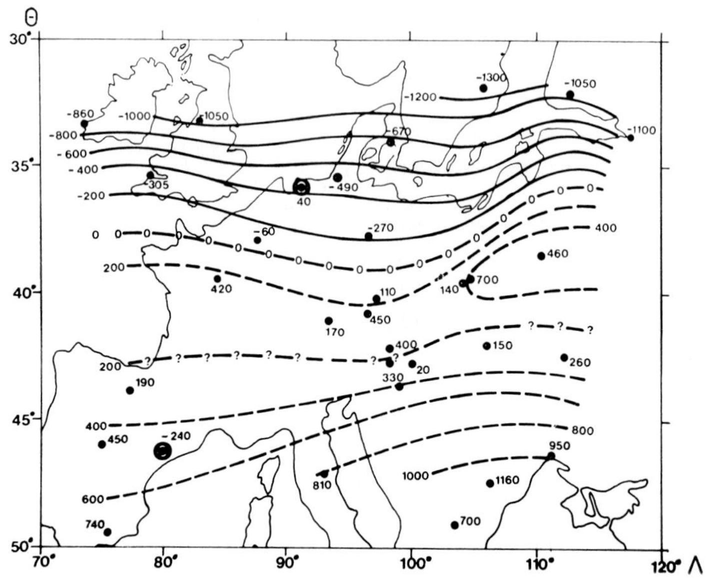

Fig. 6. Contour map of the (Z/H) deviations (multiplied by factor 10^4) over the European area for the basic 27-day harmonic.

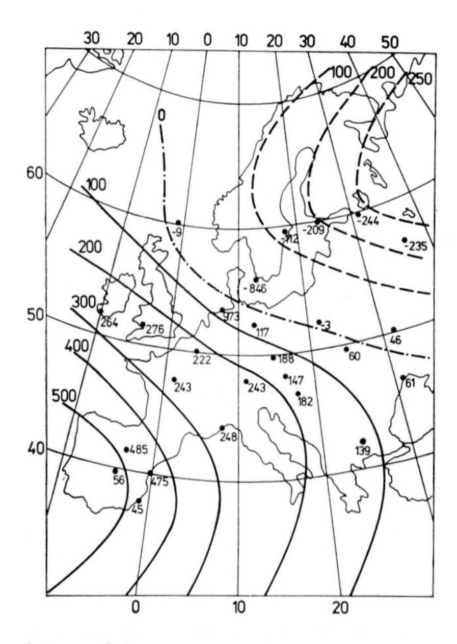

Fig. 7. Contour map of the radial component of an anomalous internal geomagnetic field induced by the 1st harmonic of the D_{st} variation (re-drawn with the omission of the numbering of observatories from the paper by FAJNBERG *et al.* (1975)). The contour values are given in nT.

Table 4. The best fitting spectral density amplitudes $A_{H,1}$, $A_{Z,1}$ (in units 1100^{-1} nT/(c·day^{-1})) and mean phases φ_H, φ_Z (in degrees).

Case	$A_{H,1}+\varepsilon_{A_H}$		$A_{Z,1}\pm\varepsilon_{A_Z}$		$\varphi_H\pm\varepsilon_{\varphi_H}$		$\varphi_Z\pm\varepsilon_{\varphi_Z}$		Period (days)
				Results of FT					
58–27	3915.2	214.8	1540.0	197.7	41.6	4.1	213.4	14.3	26.3
58–13	4239.4	185.8	1248.8	144.6	177.5	9.0	328.5	12.7	13.7
58–09	1594.7	100.9	470.6	82.6	285.2	29.6	95.6	43.0	8.98
63–27	1643.1	114.0	505.4	123.0	310.9	9.3	136.6	53.7	27.3
63–25	1426.6	76.9	490.1	91.4	87.0	5.2	259.2	33.2	25.3
63–13	1504.3	101.3	516.2	73.6	121.8	17.1	265.0	15.5	13.5
63–09	1308.0	73.0	318.9	81.7	12.0	10.0	174.2	28.4	9.06
67–27	2605.1	145.0	653.1	84.6	34.6	2.6	179.8	14.6	27.7
67–13	1641.2	91.1	440.5	78.5	313.0	5.9	116.2	23.5	13.8
67–09	992.8	73.6	260.4	59.7	64.4	11.1	292.4	55.2	9.0
				Results of FT+BNDP					
58–27	3699.6	207.2	1424.9	168.2	−304.0	7.7	−129.6	17.8	26.2
58–13	3451.2	133.4	1050.9	125.2	−154.4	14.9	−1.5	22.7	13.5
58–09	2338.4	160.2	775.9	76.6	−187.2	6.3	−15.1	19.8	9.52
63–27	1401.6	104.6	382.3	84.5	−34.5	8.0	−226.2	20.0	27.3
63–25	1252.5	89.6	418.2	63.6	−256.9	8.0	−89.2	14.0	25.3
63–13	1272.5	90.9	458.0	70.3	−214.2	7.0	−351.4	14.7	13.4
67–27	2090.2	106.3	591.1	102.7	−311.4	6.3	−164.6	14.1	27.6
67–13	1308.1	85.1	374.0	61.8	−21.1	6.3	−237.6	32.8	13.7

our zero deviation line. Certain discrepancies between the two patterns might be explained by substantial differences between the periods of the two phenomena.

4. Resistivity-Depth Profile

The best fitting amplitudes $A_{H,1}$, $A_{Z,1}$ and the phases φ_H, φ_Z obtained as the mean values of different spectral estimates for a given period are compiled in Table 4. The uncertainty in determining the best fitting amplitudes and mean values φ_H, φ_Z was expressed by standard deviations ε_H, ε_Z and ε_{φ_H}, ε_{φ_Z} which are also given in Table 4. All the values mentioned represent a basic set of data from which the internal geoelectrical structure of the Earth's mantle can be inferred.

Referring to the theoretical analysis of induction in stratified conductors by SCHMUCKER (1969, 1970a, b, 1974), we assume a uniform plane substratum of conductivity σ_c, or resistivity ρ_c, covered by a poorly conductive top layer extending from the Earth's surface to a certain depth h. It holds (SCHMUCKER, 1970a)

$$Z/H = ikc,$$

where

$$c = h + (1/2)p_c - i(1/2)p_c = z^* - i(1/2)p_c$$

is a complex-valued parameter, $k = (n+1/2)/a$ the wave number (a, the Earth's radius; n, the order of the harmonic term).

The real part of c reflects the mean depth of the internal eddy currents z^* (km), (called "depth of a perfect substitute conductor"), and the imaginary part of c indicates with $p_c=(2\pi\omega\sigma_c)^{-1/2}$ the ambient conductivity at that depth. In practical units

$$\rho^*\equiv\rho_c=(p_c/30.2)^2(1/T^*) \quad [\Omega\text{m}],$$

where T^* is the variation period in hours.

According to SCHMUCKER (1970a) the ratio (Z/H) can also be expressed by the relation

$$Z/H=iT_k,$$

where

$$T_k=\frac{1-|\bar{\iota}/\bar{e}|e^{i(\iota-\epsilon)}[(n+1)/n]}{1+|\bar{\iota}/\bar{e}|e^{i(\iota-\epsilon)}}.$$

Consequently,

$$c=(1/k)T_k=[a/(n+1/2)][\text{Re}\,(T_k)+i\,\text{Im}\,(T_k)]$$

and the parameter c can be calculated from a complex-valued surface ratio of internal to external parts of variations i.e. $|\bar{\iota}/\bar{e}|\exp[i(\iota-\epsilon)]$. The field separation can be performed by calculating amplitudes $\bar{e}$, $\bar{\iota}$ and the corresponding phases ϵ, ι from the set of amplitudes $A_{H,1}$, $A_{Z,1}$, φ_H, φ_Z (Table 4), according to formulae by RIKITAKE (1951).

Applying this approach to our actual data for three basic periods of 27, 13, and 9 days, we calculated from the mean values of $A_{H,1}$, $A_{Z,1}$, φ_H, φ_Z three relatively limited sets of couples (z, ρ^*). We also generated three sets of a large size for couples (z, ρ^*) by combining independently the signs of the standard deviations with input values $A_{H,1}\pm\varepsilon_H$, $A_{Z,1}\pm\varepsilon_Z$, $\varphi_H\pm\varepsilon_{\varphi_H}$, $\varphi_Z\pm\varepsilon_{\varphi_Z}$.

The sets were processed statistically. By testing, all groups of z and ρ^* were found to fit the normal distribution with a high coefficient of linear correlation. The mean values and the variances were then determined for specific groups of z^* and ρ^* estimates.

The determination of the wave number k is ambiguous. In our opinion, it should be specified by the relation $ka=(n+1/2)$. To substantiate it we refer to expressions for magnetic field components H_Θ, H_ϕ, H_r, (formulae (5.8), (5.9) p. 57 in SCHMUCKER, 1970a), with associated Legendre polynomials $P_n^m(\cos\Theta)$. Their asymptotic expression for $n>(1/\varepsilon)$ and $[0<\varepsilon\leq\Theta\leq\pi-\varepsilon]$ reads

$$P_n^m(\cos\Theta)\cong\frac{2}{\sqrt{\pi}}\frac{\Gamma(n+m+1)}{\Gamma(n+3/2)}\frac{1}{\sqrt{2\sin\Theta}}\cos\left[(n+1/2)\Theta-\frac{\pi}{4}+\frac{m\pi}{2}-\omega t\right].$$

Consequently, the variational field can be interpreted, at least for $n\gg1$, in terms of plane waves propagating towards the pole, or in the opposite direction, with a wave number given by the above mentioned relation. Since the alternative relation $ka=(n+1)$ is commonly used, we take into account both the possibilities in calculating z^* and ρ^* couples.

Figure 8 summarizes the depths z^* and resistivities ρ^* estimated by us from the 27-day variation and its harmonics (13- and 9-day periods) and the results derived from S_q and D_{st} variations by SCHMUCKER (1974). The wave number $k=(n+1)/a$ was assumed

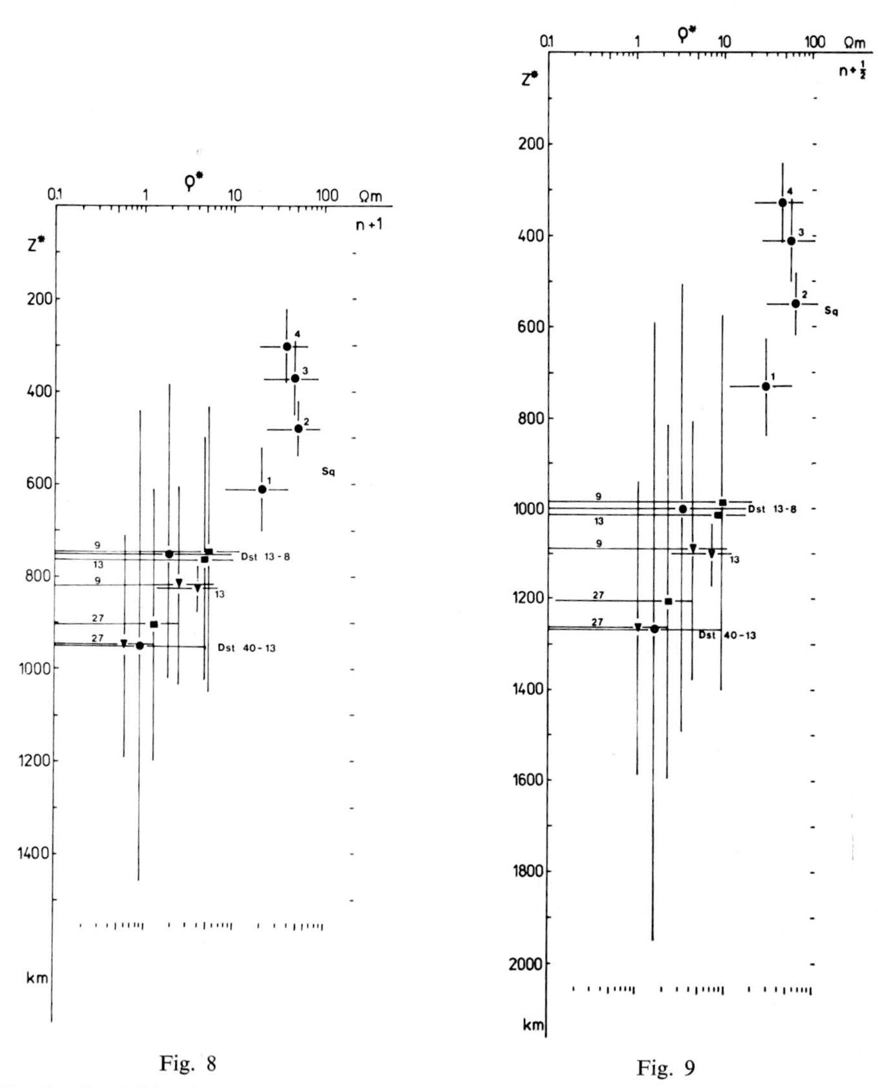

Fig. 8 Fig. 9

Fig. 8. Resistivity-depth profile of the Earth mantle suggested by SCHMUCKER (1974) and supplemented by our estimates resulting from an analysis of the 27-day variation and its 13- and 9-day harmonics. The wave number $k=(n+1)/a$ was used. Notations: full dots (●), the results by SCHMUCKER (1974); full triangles (▼), our own results obtained from mean values of both the amplitudes and phases of variations: full squares (■), our own results obtained by treating statistically the sets generated by combining the signs of the standard deviations with the input values of both the amplitudes and phases.

Fig. 9. Resistivity-depth profile of the upper mantle obtained from the same data as in Fig. 8 but with the wave number $k=(n+1/2)/a$.

in calculating the z^*, ρ^* couples.

It is apparent from the graph that our recent results fit well the resistivity-depth profile of the Earth's mantle suggested by SCHMUCKER (1974). Our mean resistivities, however, seem to be slightly shifted to higher values. They also support the existence of an important conductivity increase at the depths of about 700 km. Its steepness, according to our 27-day variation data, seems to be reduced, as compared with that suggested by S_q and D_{st} results.

The resistivity-depth profile in Fig. 9 is based on the same data as that in Fig. 8, but recalculated with the wave number $k=(n+1/2)/a$. Retaining the same general features as the previous one, the profile is extended to greater mantle depths. Consequently, the conductivity increase is shifted to the depths of about 800 and 900 km and the conductivity step is smoothed.

5. Conclusion

On analysing the geomagnetic data sets of mostly European observatories, but supplemented by results from several observatories on other continents (Table 1), we obtained the amplitude and phase characteristics of the 27-, 13-, and 9-day geomagnetic variations and their global distribution. Comparing the recent results with our previous ones, based on European data only (Pěčová *et al.*, 1977a, b), we can conclude:

1) Assuming in the 1st order approximation a purely zonal distribution of the 27-day variation and of its harmonics, we find that there are no substantial differences between the results of both analyses concerning the amplitudes $A_{H,1}$, $A_{Z,1}$, phase difference $\varphi_Z-\varphi_H$ and all related characteristics involving the response function Q_1, the amplitudes $\bar{e}$, $\bar{\iota}$ and the phases ϵ, ι of both the internal and external parts of the variation potential. Consequently, the European continent does not differ substantially in its internal geoelectrical structure from that of the entire globe.

2) There is a satisfactory agreement between the results of both the recent and the previous trend analyses. The spatial distribution of both the H and Z components displayed in form of contour maps in Figs. 4, 5 retain their previous characteristics. The H component depends mostly on the geomagnetic latitude, while the Z component, specifically in accordance with the 2nd order trend surface, shows the dependence on both geomagnetic co-ordinates.

3) The distribution of deviations, defined for the (Z/H) ratio and analysed previously, remains essentially unchanged over the European region. Large areas with negative and positive deviations are separated by a zero-deviation line traversing the northern part of Central Europe in an east-west direction. The deviation pattern is displayed in Fig. 6 for the basic 27-day harmonic. The sector of negative deviations covering mainly the northern part of Europe might be a region with increased electrical conductivity at depths of about 1,000 km.

Using the theoretical basis of electromagnetic induction in stratified conductors by Schmucker (1970a), we estimated the depths of a perfect substitute conductor and the ambient resistivity at those depths from the characteristics of the 27-, 13-, and 9-day variations. The present estimates fit well the Schmucker's resistivity-depth profile of the Earth's mantle based on the results of analysing the D_{st} variations within the range of periods between 40 and 8 days.

REFERENCES

BANKS, R. J., Geomagnetic variations of the electrical conductivity of the upper mantle, *Geophys. J. R. Astr. Soc.*, **17**, 457–487, 1969.

FAJNBERG, E. B., V. G. DUBROVSKIJ, and L. L. LAGUTINSKAYA, Okeanitcheskij effekt v pole *D-st* variacij, in *Analiz Prostranstvenno-Vremennoj Struktury Geomagnitnogo Polia*, pp. 130–152, Nauka, Moskva, 1975.

PĚČOVÁ, J., O. PRAUS, and K. PĚČ, Long period variations of the geomagnetic field and their spatial distribution in Europe, *Studia geoph. et geod.*, **21**, 148–158, 1977a.

PĚČOVÁ, J., O. PRAUS, and K. PĚČ, Spatial distribution of long period geomagnetic variations over European area, *Acta Geodaet., Geophys. et Montanist. Acad. Sci. Hung.*, **T. 12**, 407–415, 1977b.

RIKITAKE, T., Electromagnetic induction within the Earth and its relation to electrical state of the Earth's interior, Part III, *Bull. Earthq. Res. Inst.*, **29**, 61–69, 1951.

SCHMUCKER, U., *Regionale Unterschiede im inneren Anteil des S_q-Ganges, Protokoll über des Kolloquium "Erdmagnetische Tiefensondierung"*, pp. 181–189, Reinhausen bei Göttingen, 4.–6. März, 1969.

SCHMUCKER, U., *Anomalies of Geomagnetic Variations in the Southwestern United States*, pp. 55–76, Univ. of California, Berkeley-L. Angeles-London, 1970a.

SCHMUCKER, U., An introduction to induction anomalies, *J. Geomag. Geoelectr.*, **22**, 9–33, 1970b.

SCHMUCKER, U., Erdmagnetische Tiefensondierung mit langperiodischen Variationen, Protokoll über das Kolloquium "Erdmagnetische Teifensondierung", Grafrath/Bayern, 11.–13. März, pp. 313–342, 1974.

An Interpretation of the Induction Arrows at Indian Stations*

B. J. SRIVASTAVA and H. ABBAS

National Geophysical Research Institute, Hyderabad, India

(Received September 25, 1979; Revised February 10, 1980)

Induction arrows (Wiese vectors) for night-time SSCs and Bays are determined and discussed along with geological and other geophysical data at eight Indian magnetic observatories, from Sabhawala in the Himalayan foothills to Trivandrum on the seacoast in the extreme south.

A medium-sized induction arrow at Sabhawala pointing northwards indicates that the region of higher subsurface conductivity lies to the south of Sabhawala, possibly along the Aravalli Hills, and not beneath the Kashmir Himalaya, where the conductive upper mantle appears to be depressed. East-south-east pointing arrows at Jaipur and Ujjain are again indicative of a high electrical subsurface conductivity on the west along the Aravalli Hills and the Cambay region, where high heatflow and gravity values have also been reported, and the upper mantle appears to be elevated.

At Alibag, a medium-sized induction arrow points eastward, with the conductive seawater lying on the west. Hyderabad is the only station where small induction arrows (small Z variations), characteristic of an inland station, are observed. Negative Z variations for positive H variations along the Alibag-Hyderabad-Kalingapatnam profile, become positive about 100 km east of Hyderabad due to ocean effect from the east coast.

Unusually large induction arrows (very large positive Z variations) in the extreme south Peninsular tip (largest one at Trivandrum), so close to the dip equator, are indicative of very strong induced current concentrations near and along this coast. These oceanic induced currents causing the anomalous geomagnetic variations in Peninsular India, appear to concentrate along the coastline and the continental shelf, and flow from north to south along the east coast. The currents concentrate further as they pass through the narrow Palk Strait and the Gulf of Manar between India and Sri Lanka on to Kanyakumari and the Trivandrum coast, and turn northwards along the west coast due to the obstruction provided by the volcanic ridge of Lakshadweep, Minicoy and Amindivi islands lying to the west in the Arabian Sea, about 300 km off the Trivandrum coast. The current concentration would turn westwards near Calicut (11°N) and pass north of these islands in a diffuse form. Another current concentration is expected to flow along the Alibag west coast from north to south (so as to explain the reversed coast effect at Alibag), and turn westwards near Calicut to merge with the Bay of Bengal current system. There is no need to postulate a couductor and current channelling beneath the sea near the Trivandrum coast without any supporting geological evidence. There could be a channelling of the induced currents at the interface of the conductive upper mantle beneath the ocean and the less conductive upper mantle

* Presented at the Fourth Workshop on EM Induction in the Earth and Moon, Murnau, September 1978.

beneath the Peninsular India (100–200 km), in a step-structure around the coast, which will account for the induction anomalies observed in long-period variations like *Sq*.

1. Introduction

The geomagnetic induction anomalies in India were discovered more than a decade ago (SRIVASTAVA, 1966; SRIVASTAVA and SANKER NARAYAN, 1967, 1969). This was followed by more intensive investigations for the delineation of the anomalies and also for better appreciation of the oceanic and subsurface causes of the amonalies identified mostly from the records of the permanent magnetic observatories in Peninsular India (SRIVASTAVA, 1970, 1977a; SRIVASTAVA *et al.*, 1974a, b; NITYANANDA *et al.*, 1977). We now have in the longitude zone 72°E–80°E in India, a chain of ten magnetic observatories from Gulmarg in the Kashmir Himalaya (very recently established by the Indian Institute of Geomagnetism) and Sabhawala (operated by the Survey of India since 1963) in the Himalayan foothills and close to the *Sq* focus in the north, down to Trivandrum on the seacoast in the extreme south. One more observatory has been commissioned recently at Shillong in the north-eastern region of India, near the latitude of Jaipur. The four southern stations in the Peninsular tip are located very close to the dip equator (Trivandrum being to its south), and are directly under the influence of the day-time equatorial electrojet. As is well known, the electrojet provides a very intense and non-uniform type of electromagnetic inducing field in this region during day time. It was therefore decided to examine only night-time rapid geomagnetic variations like SSCs and bays, whose external fields are fairly uniform in these latitudes during the night hours when the equatorial electrojet is absent.

The induction arrows (Wiese vectors) at eight of the eleven Indian stations have been computed and are given in this communication, following WIESE (1962) and UNTIEDT (1970). An interpretation of these arrows in terms of the local geology, oceanic induction effects and undulations of the conductive upper mantle has also been attempted.

2. Data and Analysis

Figure 1 shows the locations of the eleven magnetic observatories in India currently in operation. The induction arrows (Wiese vectors for night-time SSCs and bays are also shown separately at the eight stations (indicated by solid circles). The choice of the data used in this investigation was restricted by the availability of the data to the authors. The night-time SSC data comprising the amplitudes in *H*, *D*, and *Z* components were mainly taken from the "Indian Magnetic Data" volumes for the periods 1969–1974, and "Solar-Geophysical Data" volumes for the period 1975–1977. Data on selected prominent bay events (*H*, *D*, and *Z* amplitudes) were compiled from the available copies of the relevant records of the stations for data taken during 1967, 1974, and 1977. The *D* amplitudes were appropriately converted to nT units. The *H* and *Z* amplitudes were already given in nT units. In the case of Hyderabad, however, all the prominent night-time SSCs and bays recorded during 1969–1977 were used, while for Kodaikanal only a few selected bay events of 1967 were available. Relatively few events were available

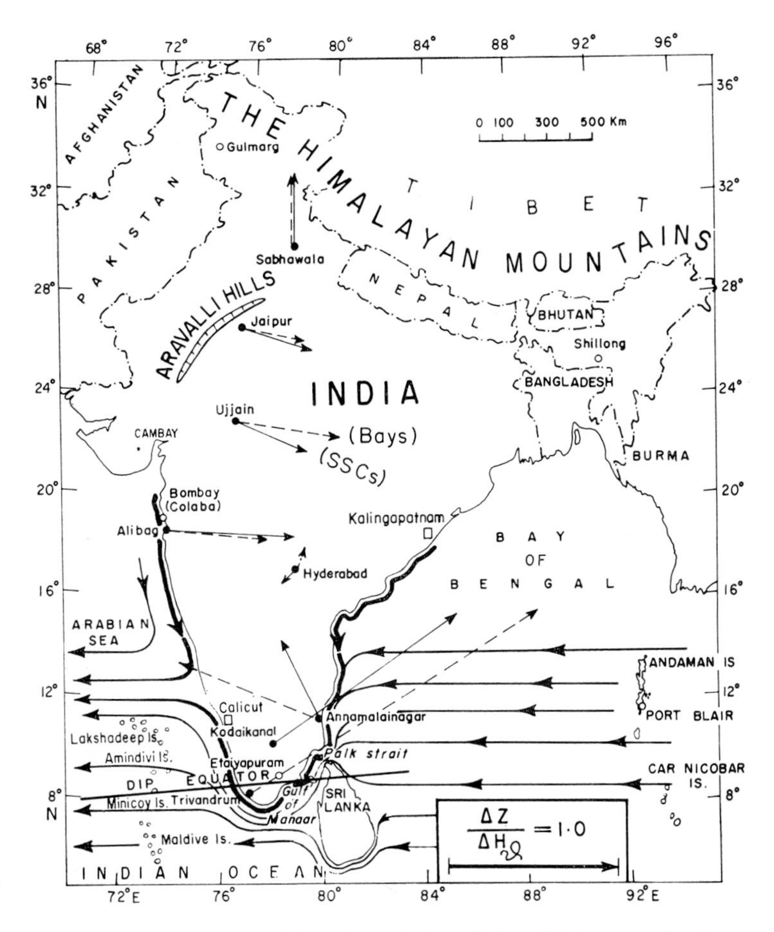

Fig. 1. Map of India showing the locations of the existing eleven magnetic observatories. At the eight observatories shown with solid circles, the induction arrows (Wiese vectors) for night-time SSC and bay events are also given. Possible oceanic induced electric currents during these geomagnetic variations, and their distortions near the Peninsular coastline are also roughly indicated.

Table 1. Coordinates of the eight magnetic observatories used in this study.

Observatory	Geographic		Geomagnetic	
	Latitude	Longitude	Latitude	Longitude
Sabhawala	30°22 N	77°48 E	+20.8°	149.8°
Jaipur	26 55 N	75 48 E	+17.3	147.4
Ujjain	23 11 N	75 47 E	+13.5	147.0
Alibag	18 38 N	72 52 E	+ 9.5	143.6
Hyderabad	17 25 N	78 33 E	+ 7.6	148.9
Annamalainagar	11 22 N	79 41 E	+ 1.4	149.4
Kodaikanal	10 14 N	77 28 E	+ 0.6	147.1
Trivandrum	8 29 N	76 57 E	− 1.2	146.4

for Jaipur and Ujjain. H amplitudes of all the events used in this study were greater than 10 nT. Table 1 gives the coordinates of these stations.

In Fig. 2 are reproduced the H, D, and Z records of a recent typical night-time SSC taken at the Indian stations, while Fig. 3 shows a typical bay event.

Following UNTIEDT (1970), $\Delta D/\Delta H$ and $\Delta Z/\Delta H$ ratios were computed for the individual SSC and bay events at each of the eight stations. These were then plotted as x and y coordinates. From the cluster of points so obtained, the lines of best fit were visually drawn (Fig. 4). From the intercepts of these lines on the y and x axes, the con-

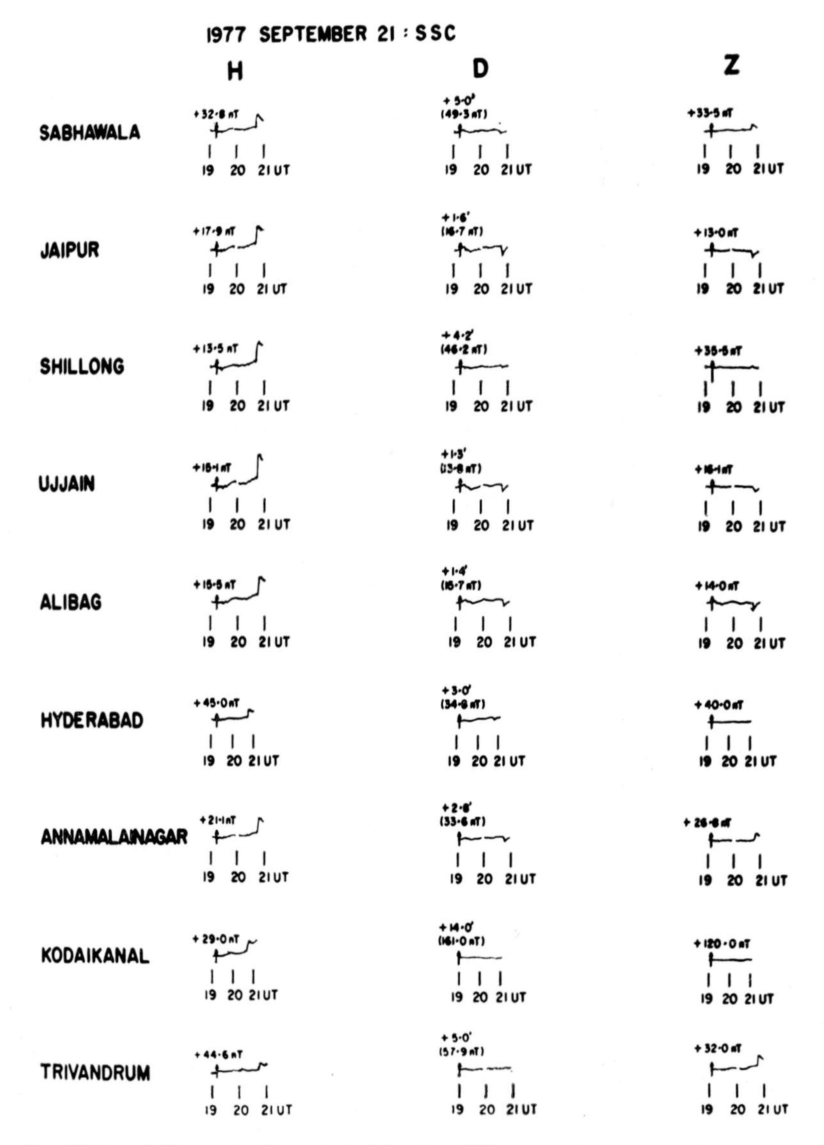

Fig. 2. H, D, and Z records of a typical night-time SSC event taken at the Indian stations on September 21, 1977, at 2043 UT.

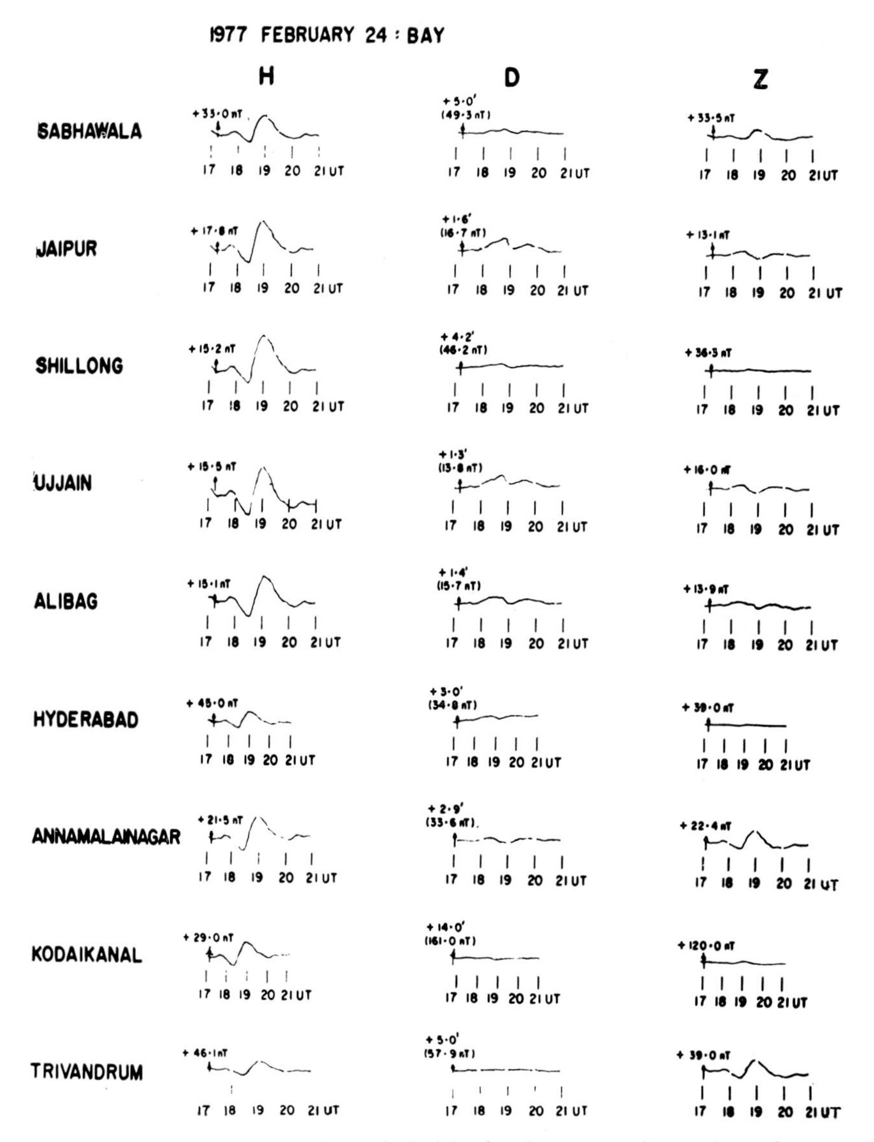

Fig. 3. *H*, *D*, and *Z* records of a typical night-time bay event taken at the Indian stations on February 24, 1977, at 1830 UT.

stants *a* and *b* were determined, using the equation

$$\Delta Z = a\Delta H + b\Delta D \tag{1}$$

where ΔZ, ΔH, and ΔD represent the amplitudes in the respective components in nT for the events considered.

With *a* as *y* coordinate and *b* as *x* coordinate, the induction vectors, $c=(a^2+b^2)^{1/2}$, were determined and drawn from the coordinate origin for each of the two sets of events (SSCs and bays) at each of the eight stations (Figs. 1 and 4). In the case of bays at

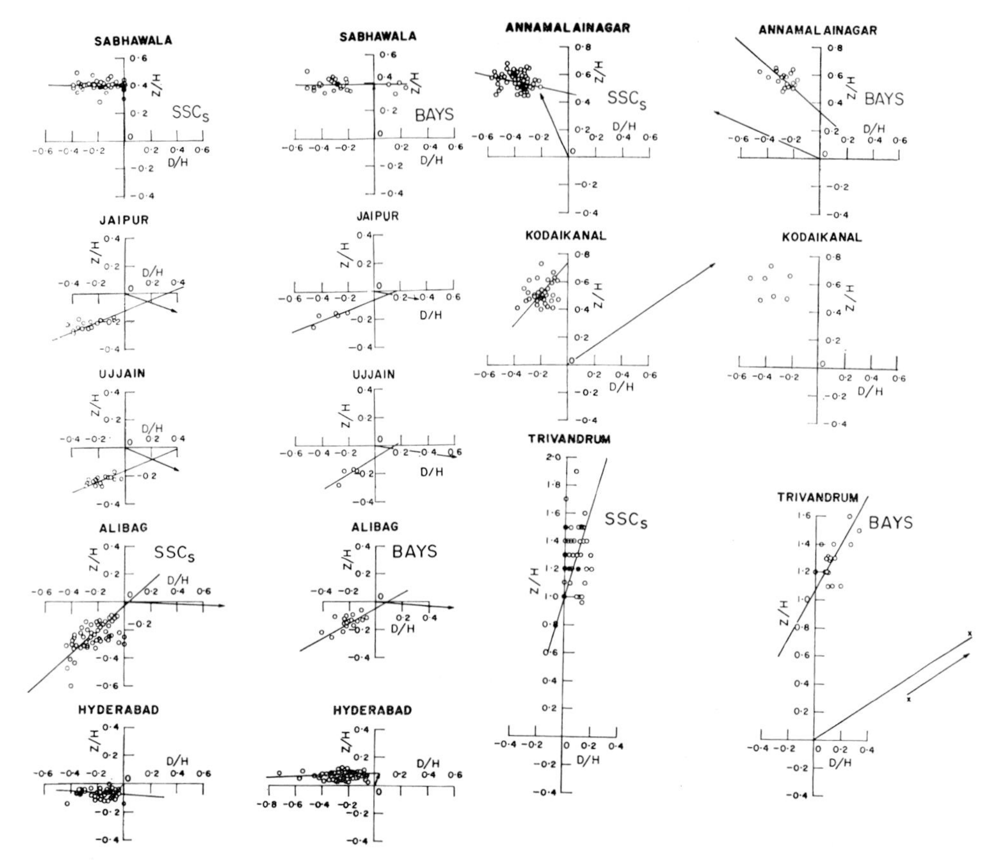

Fig. 4. Plots of $\Delta Z/\Delta H$ versus $\Delta D/\Delta H$ of night-time SSCs and bays at the eight Indian stations in the longitude zone 72°E–80°E. The best-fitting lines and the induction arrows are also shown.

Kodaikanal and SSCs at Trivandrum, the induction arrows could not be determined as is clear from the scatter diagram (Fig. 1).

3. Discussion and Interpretation

3.1 The northern region

The close resemblance of H and Z traces for both SSC and bay events at Sabhawala (Figs. 2 and 3), readily suggests that the anomalous internal Z amplitudes arise from induction by the external H field in a nearly east-west subterranean conductor. This point is fully borne out by the medium-sized induction arrows at Sabhawala pointing northwards (Fig. 1). It would further indicate that the region of higher subsurface conductivity that lies to the south of Sabhawala, may be beneath the southern Siwaliks (foothills of the Himalaya) or along the strike of the Aravalli Hills, but not beneath the middle Himalaya in Kashmir.

According the DEWEY and BIRD (1970) the different mountain belts of the world were formed as a result of the collision of lithospheric plates of different kinds. They found that if a trench develops near the continental margin to consume lithosphere from the oceanic side, a mountain belt (Cordilleran type) grows dominantly by thermal mechanisms. Such mountain belts as exist in Southern British Columbia, Canada, are characterized by paired metamorphic belts (blueschist on the oceanic side and high temperature on the continental side) and divergent thrusting from the high temperature volcanic axis. NIBLETT *et al.* (1969) discussed the Cordillera anomaly in geomagnetic veriations arising from such a conductivity structure beneath the mountain belt. SCHMUCKER (1969) reported similar results from the southwest United States. SCHMUCKER (1969) found a highly conductive structure beneath the Andean mountains in South America, from magnetometer observations at a network of temporary stations across the Andes.

The Himalayan mountains have formed as a result of collision between two continental plates. Such mountain belts will be large and will not have paired metamorphic belts. They have a single dominant direction of thrusting away from the site of the trench over the underthrust plate. The root of such collisional mountain belts is sialic as opposed to the root of the Cordilleran type which is basic due to volcanism and high temperature metamorphic axis (DEWEY and BIRD, 1970). It is quite possible that the conductive upper mantle beneath the Himalaya is sufficiently depressed and is overlain by a thick crust of 60 to 70 km, while it is uplifted in the plains and the Aravalli Hills to 100–200 km, south of the Himalaya, thereby providing a step-structure in the upper mantle, which would appropriately channelize the induced eddy currents to give rise to the type of induction anomalies observed at Sabhawala (large positive Z variations, as opposed to negative Z variations at Jaipur, Ujjain, Alibag and small negative changes at Hyderabad, for positive H variations at all the stations). LAW and RIDDIHOUGH (1971) also noted a relationship between inland geomagnetic variation anomalies and tectonic features like the Himalaya, around the world, and associated them with plate margins.

3.2 The western and central region

The medium-sized induction arrows at Jaipur and Ujjain pointing east-south-eastwards (Fig. 1) are again indicative of a high electrical subsurface conductivity on the west along the strike of the Aravalli Hills and the Cambay region. SRIVASTAVA (1977b, c) has already pointed out that the conductive upper mantle in this region is uplifted. It is further supported by observations of high heat flow values (3 μcal/cm^{2}sec) and high gravity values along the Cambay coast.

At Alibag, a medium-sized induction arrow points eastward, with the highly conductive seawater ($\sigma \sim 4\ \Omega^{-1}\mathrm{m}^{-1}$) lying to the west. The induced current concentration both in the seawater as well as in the upper mantle step-structure along the coastline will have to flow from north to south as depicted in Fig. 1. This will explain the reversed coast effect observed at Alibag for events like SSCs and bays with larger negative Z amplitudes near the coast for positive H variations (SRIVASTAVA *et al.*, 1974a) and larger $Sq(Z)$ values at Alibag as compared to Hyderabad (SRIVASTAVA, 1970). The current concentration is expected to turn westwards from the coast near Calicut (SRIVASTAVA and SANKER NARAYAN, 1969). Hyderabad is the only station where the induction arrows are found to be small with very different orientations for SSCs and bays. This is characteristic of an inland station (as at Alice Spring in Australia, PARKINSON, 1962), where very small Z variations are observed—negative Z amplitudes for SSCs and positive for bays, for positive H amplitudes. SRIVASTAVA *et al.* (1974b) found that along the Alibag-Hyderabad-Kalingapatnam profile, the negative Z variations for positive H variations and negative D variations, become positive at a distance of about 100 km east of Hyderabad due to ocean effect from the east coast. Since larger $Sq(Z)$ values are also observed near the east coast (Kalingapatnam), the induced electric currents along the east coast will again concentrate along and near the coastline in the seawater as well as in the step-like structure in the conductive upper mantle, and flow from north to south (the upper mantle being elevated under the sea, as was pointed out by PARKINSON (1962, 1963)). Hyderabad turns out to be an ideal inland low-latitude station, outside the influence of the day-time equatorial electrojet and the oceanic induced electric currents.

3.3 The southern electrojet region

The region of the southern Peninsular tip of India containing the dip equator and the day-time equatorial electrojet, reveals one of the most complex electromagnetic induction anomalies in the world (SRIVASTAVA, 1977b, c). The induction arrows are very large at the two coastal stations of Annamalainagar and Trivandrum as well as the inland station of Kodaikanal, and are indicative of the higher oceanic conductivity (Fig. 1). The anomaly is severest at Trivandrum where the induction arrow exceeds 1. Earlier studies have shown (SRIVASTAVA and SANKER NARAYAN, 1967; SRIVASTAVA, 1970, 1977c) that the induction anomalies in this region are observed in both short-period (SSC, bay) and long-period (Sq, L) variations at permanent and temporary stations. It has been demonstrated that not only the Z variations are anomalously large in the electrojet region but the H and D variations are also anomalous at Annamalainagar and Trivandrum (SRIVASTAVA and SANKER NARAYAN, 1969; NITYANANDA *et al.*, 1977). Separation of the variation fields into external and internal parts has not yet been carried out, for want of a close network of stations. The large induction arrows (Fig. 1) so close to the dip equator are indicative of very strong induced current concentrations near and along this coast. As already suggested by SRIVASTAVA and SANKER NARAYAN (1969), the induced currents in the seawater as well as in the upper mantle in this region for night-time bay events would be mainly in the E–W direction, the external inducing field being the H component of the variations (Figs. 2 and 3). These oceanic induced currents appear to concentrate along the coastline and the continental shelf due to conductivity contrast, and flow from nearly north to south along the east coast. The currents would concentrate further as they pass through the narrow Palk Strait and the Gulf of Manar between India and Sri Lanka on to Kanyakumari and the Trivandrum coast. They turn northwards along the west coast due to the obstacle provided by the volcanic ridge of Lakshadweep, Minicoy, and Amindivi islands lying to the west in the Arabian Sea, about 300 km off the Trivandrum coast (Fig. 1). They would then turn westwards near Calicut (11°N) and pass north of these islands in a diffuse form, and merge with the Arabian Sea current concentration along the Alibag west coast, which too turns westwards to form the general pattern of the induced currents in the seawater in the E–W direction.

It is also expected that there would be a similar concentration of induced currents at the interface of the conductive upper mantle beneath the ocean and the less conductive upper mantle beneath the Peninsular India at a depth of 100–200 km, in a conductive step-structure around the coast, in order to account for the induction anomalies observed in geomagnetic bays and longer period variations like *Sq*.

There is no need to postulate an isolated conductor with current channelling therein beneath the sea near the Trivandrum coast (Singh, 1977), without supporting geological evidence as also shown by the model calculations of Takeda and Maeda (1979). The model experiments carried out by Papamastorakis (1975) simulating induction in the Bay of Bengal and the Arabain Sea, will have to be refined further.

The complex nature of the induction problem in the electrojet region in India will have to be fully understood, and appropriate allowance made for anomalous internal variations in the region, before computing the parameters of the electrojet.

4. Concluding Remarks

The geomagnetic induction anomalies identified at the various stations from Sabhawala to Trivandrum, with the help of the induction arrows suggest undulations of the conductive upper mantle (100–200 km) or the asthenosphere. A depression of the mantle beneath the Kashmir Himalaya, elevation in the west central India including the Cambay region and the Aravalli Hills, and again a depression in the electrojet region followed by an upliftment beneath the Indian Ocean in the south, apparently matches the available geomagnetic observations.

The complex problems involved in these studies cannot be resolved unless magnetometer array experiments are carried out in each of the above regions, and the internal anomalous data appropriately interpreted along with other geophysical and geological data from these regions.

The authors are grateful to Dr. A. Roy, Director, National Geophysical Research Institute, Hyderabad, for permission to present this paper at the Murnau Workshop on EM Induction. Thanks are also due to the Directors of the Indian Institute of Geomagnetism, Indian Institute of Astrophysics, Geodetic and Research Branch of the Survey of India, for readily supplying copies of the relevant magnetic records from their observatories, used in this investigation.

REFERENCES

Dewey, J. F. and J. M. Bird, Mountain belts and the new global tectonics, *J. Geophys. Res.*, **75**, 2625–2647, 1970.

Law, L. K. and R. P. Riddihough, A geographical relation between geomagnetic variation anomalies and tectonics, *Can. J. Earth Sci.*, **8**, 1094–1106, 1971.

Niblett, E. R., B. Caner, and K. Whitham, Electrical conductivity anomalies in the mantle and crust in Canada, in *The Application of Modern Physics to the Earth and Planetary Interiors*, edited by S. K. Runcorn, pp. 155–172, Wiley-Inter Science Publications, London, 1969.

Nityananda, N., A. K. Agarwal, and B. P. Singh, Induction at short period in the horizontal field variations in Indian Peninsula, *Phys. Earth Planet. Inter.*, **15**, 5–9, 1977.

Papamastorakis, I., Max Planck Institut Fur Physik and Astrophysik, Ph. D. Thesis, Munchen, F.R.G., 1975.

Parkinson, W. D., The influence of the continents and oceans on geomagnetic variations. *Geophys. J.*

R. Astr. Soc., **6**, 441–449, 1962.

PARKINSON, W. D., Conductivity anomalies in Australia and the ocean effect, *J. Geomag. Geoelectr.*, **15**, 222–226, 1963.

SCHMUCKER, U., Conductivity anomalies with special reference to the Andes, in *The Application of Modern Physics to the Earth and Planetary Interiors*, edited by S. K. Runcorn, pp. 125–138, Wiley-Inter Science Publications, London, 1969.

SINGH, B. P., Subsurface structure and the equatorial electrojet, Proc. Workshop Equat. Electrojet and Associated Phenomena, PRL, Ahmedabad (India), 25–29 Oct. 1977, pp. 39–50, 1977.

SRIVASTAVA, B. J., The Hyderabad geomagnetic data for the year 1965— a comparison against the Alibag data for 1945 and 1958, *Bull. NGRI*, **4**, 119–122, 1966.

SRIVASTAVA, B. J., Intercomparison of magnetometer measurements in India and its implications, Proc. Symp. Problems Equatorial Electrojet, Ahmedabad, 8 August 1970, Paper No. 11, 1970.

SRIVASTAVA, B. J., Comments on the paper entitled "Variations in *Z* during magnetic storms at Alibag, Annamalainagar and Trivandrum", *Inidan J. Radio Space Phys.*, **6**, 205–207, 1977a.

SRIVASTAVA, B. J., Electromagnetic Induction anomalies in relation to the continental margins of India, IAGA: IWG Symp. SEI IAGA/IAMAP Joint Assembly, Seattle, U.S.A., 22 Aug.–3 Sept. 1977, E.O.S., 58, 782 (GA 675), 1977b.

SRIVASTAVA, B. J., Anomalous geomagnetic variations at electrojet stations in Idian due to coastal and subsurface causes, Proc. Workshop Equatorial Electrojet and Associated Phenomena P.R.L., Ahmedabad (India), 25–29 October 1977, pp. 28–37, 1977c.

SRIVASTAVA, B. J. and P. V. SANKER NARAYAN, Anomaly of geomagnetic time-variations observed on the Peninsular India, Proc. Symp. Upper Mantle Project, Hyderabad, January 1967, pp. 165–174, 1967.

SRIVASTAVA, B. J. and P. V. SANKER NARAYAN, Anomalous geomagnetic variations during night-time bays at Indian Observatories, Proc. 3rd Intl. Symp. Equatorial Aeronomy, Ahmedabad (India), 3–8 February 1969, pp. 468–473, 1969.

SRIVASTAVA, B. J., D. S. BHASKARA RAO, and S. N. PRASAD, Geomagnetic variation anomalies in the Koyna and the Bhadrachalam seismic areas in Peninsular India, *J. Geomag. Geoelectr.*, **26**, 247–255, 1974a.

SRIVASTAVA, B. J., D. S. BHASKARA RAO, and S. N. PRASAD, Geomagnetic induction anomalies along the Hyderabad-Kalingapatnam profile, *Tectonophysics*, **24**, 343–350, 1974b.

TAKEDA, M. and H. MAEDA, Effect of the coastline configuration of South Indian and Sri Lanka regions on the induced field at short period, *J.Geophys.*, **45**, 209–218, 1979.

UNTIEDT, J., Conductivity anomalies in central and southern Europe, *J. Geomag. Geoelectr.*, **22**, 131–149, 1970.

WIESE, H., Geomagnetische Tiefentellurik. Teil II: Die Streichrichtung der Untergrundstrukturen des elektrischen Widerstandes, erschlossen aus geomagnetischen variationen, *Geofis. Pura e Appl.*, **52**, 83–103, 1962,

AEPS Vol. 1

Hard cover edition to Journal of Geomagnetism and Geoelectricity Vol. 29, No. 4, 1977

Proceedings of AGU 1976 Fall Annual Meeting, December 1976, San Francisco

ORIGIN OF THERMOREMANENT MAGNETIZATION

Edited by David J. DUNLOP

Contents TRM and Its Variation with Grain Size: A Review (R. DAY)/Single Domain Oxide Particles as a Source of Thermoremanent Magnetization (M.E. EVANS) / Domain Structure of Titanomagnetities and Its Variation with Temperature (H.C. SOFFEL) / The Demagnetization Field of Multidomain Grains (R.T. MERRILL) / The Hunting of the 'Psark' (D.J. DUNLOP) / On the Origin of Stable Remanence in Pseudo-Single Domain Grains (S.K. BANERJEE) / The Preparation, Characterization and Magnetic Properties of Synthetic Analogues of Some Carriers of the Palaeomagnetic Record (J.B. O'DONOVAN and W. O'REILLY) / Reduction of Hematite to Magnetite under Natural and Laboratory Conditions (P.N. SHIVE and J.F. DIEHL) / Characteristics of First Order Shock Induced Magnetic Transitions in Iron and Discrimination from TRM (P. WASILEWSKI) / The Thermoremanence Hypothesis and the Origin of Magnetization in Iron Meteorites (A. BRECHER and L. ALBRIGHT) / Thermal Overprinting of Natural Remanent Magnetization and K/Ar Ages in Metamorphic Rocks (K.L. BUCHAN, G.W. BERGER, M.O. MCWILLIAMS, D. YORK, and D.J. DUNLOP) / Does TRM Occur in Oceanic Layer 2 Basalts? (J.M. HALL) / The Effects of Alteration on the Natural Remanent Magnetization of Three Ophiolite Complexes: Possible Implications for the Oceanic Crust (S. LEVI and S.K. BANERJEE)

AEPS Vol. 2

Hard cover edition to Journal of Physics of the Earth Vol. 25, Supplement, 1977 (Not included in regular issues)

Proceedings of the U.S.-Japan Seminar on Theoretical and Experimental Investigations of Earthquake Precursors

EARTHQUAKE PRECURSORS

Edited by C. KISSLINGER and Z. SUZUKI

Contents Earthquake Prediction-Related Research at the Seismological Laboratory, California Institute of Technology, 1974–1976 (J.H. WHITCOMB)/Research on Earthquake Prediction and Related Areas at Columbia University (L.R. SYKES)/Seismic Activities and Crustal Movements the Yamasaki Fault and Surrounding Regions in the Southwest Japan (K. OIKE) / The New Madrid Seismic Zone as a Laboratory for Earthquake Prediction Research (B.J. MITCHELL, W. STAUDER, and C.C. CHENG) / Anomalous Crustal Activity in the Izu Peninsula, Central Honshu (K. TSUMURA) / Recent Seismometrical Works in Japan (S. SUYEHIRO, M. ICHIKAWA, and K. TSUMURA) / Quiet and Violence in Horizontal Movement of the Crust (T. HARADA) / Anomalous Seismic Activity and Earthquake Prediction (H. SEKIYA) / Seismic Activity in the Northeastern Japan Arc (A. TAKAGI, A. HASEGAWA, and N. UMINO) / Observations of Changes in Seismic Wave Velocity in South Kanto District, South of Tokyo, by the Explosion-Seismic Method (T. KAKIMI and I. HASEGAWA) / Some Precursors Prior to Recent Great Earthquakes along the Nankai Trough (H. SATO) / Possibility of Temporal Variations in Earth Tidal Strain Amplitudes Associated with Major Earthquakes (T. MIKUMO, M. KATO, H. DOI, Y. WADA, T. TANAKA, R. SHICHI, and A. YAMAMOTO) / Gravity Changes Associated with Seismic Activities (Y. HAGIWARA) / Geomagnetism in Relation to Tectonic Activities of the Earth's Crust in Japan (N. SUMITOMO) / Precursory and Coseismic Changes in Ground Resistivity (T. RIKITAKE and Y. YAMAZAKI) / Geochemistry as a Tool for Earthquake Prediction (H. WAKITA) / Recent Laboratory Studies of Earthquake Mechanics and Prediction (W.F. BRACE) / Dilatancy of Rocks under General Triaxial Stress States with Special Reference to Earthquake Precursors (K. MOGI) / Possibility of a Great Earthquake in the Tokai District, Central Japan (T. UTSU) / Depth Constraints on Dilatancy Induced Velocity Anomalies (K.W. WINKER and A. NUR) / Seismological Precursors to a Magnitude 5 Earthquake in the Central Aleutian Islands (E.R. ENGDAHL and C. KISSLINGER) / Estimation of Future Destructive Earthquakes from Active Faults on Land in Japan (T. MATSUDA) / Some Problems in the Prediction of the Nemuro-oki Earthquake (K. ABE) / Responses to Earthquake Prediction in Kawasaki City, Japan in 1974 (H. OHTA and K. ABE) / Socioeconomic and Political Consequences of Earthquake Prediction (J.E. HAAS and D.S. MILETI)

AEPS Vol. 3

Proceedings of the U.S.-Japan Seminar on Rare Gas Abundance and Isotopic Constraints on the Origin and Evolution of the Earth's Atmosphere

TERRESTRIAL RARE GASES

Edited by E.C. ALEXANDER, Jr. and M. OZIMA

Contents *EXPERIMENTAL STUDIES* A Mantle Helium Component in Circum-Pacific Volcanic Gases: Hakone, the Marianas, and Mt. Lassen (H. CRAIG, J.E. LUPTON, and Y. HORIBE) / Nitrogen to Argon Ratio in Volcanic Gases (S. MATSUO, M. SUZUKI, and Y. MIZUTANI) / Rare Gas Abundance Pattern of Fumarolic Gases in Japanese Volcanic Areas (O. MATSUBAYASHI, S. MATSUO, I. KANEOKA, and M. OZIMA) / A Review: Some Recent Advances in Isotope Geochemistry of Light Rare Gases (I.N. TOLSTIKHIN) / Abundances and Isotopic Compositions of Rare Gases in Granites and Thucholites (P.K. KURODA and R.D. SHERRILL) / Rare Gas Isotopic Compositions in Diamonds (N. TAKAOKA and M. OZIMA) / Rare Gases in Mantle-Derived Rocks and Minerals (I. KANEOKA, N. TAKAOKA, and K. AOKI) / A Comparison of Terrestrial and Meteoritic Noble Gases (O.K. MANUEL) / The Composition and History of the Martian Atmosphere (T. OWEN) *THEORETICAL STUDIES* Nuclear Components in the Atmosphere (T.J. BERNATOWICZ and F.A. PODOSEK) / Trapped Xenon and Cosmic-Ray Effects in Meteorites, in Lunar Sample, and in the Earth's Materials (K. SAKAMOTO) / Classification and Generation of Terrestrial Rare Gases (K. SAITO) / Earth-Atmosphere Evolution Model Based on Ar Isotopic Data (Y. HAMANO and M. OZIMA) / Terrestrial Potassium and Argon Abundances as Limits to Models of Atmospheric Evolution (D.E. FISHER) / On the Ambient Mantle $^{4}He/^{40}Ar$ Ratio and the Coherent Model of Degassing of the Earth (D.W. SCHWARTMAN) / Earth Degassing Models, and the Heterogenous vs. Homogeneous Mantle (R. HART and L. HOGAN) / Lead Isotope Constraints on the Early History of the Earth (R.D. RUSSELL) / Matter Accretion into the Solar System (S. HAYAKAWA)

AEPS Vol. 4

Hard cover edition to Journal of Geomagnetism and Geoelectricity Vol. 30, Nos. 3 and 4, 1978

Proceedings of IAGA/IAMAP Joint Assembly, August 1977, Seattle, Washington

AURORAL PROCESSES

Edited by C.T. RUSSELL

Contents *TIMING OF SUBSTORM EVENTS* Pi 2 Micropulsations as Indicators of Substorm Onsets and Intensifications (G. Rostoker and J.V. Olson) / The Use of Ground Magnetograms to Time the Onset of Magnetospheric Substorms (R.L. McPherron) / Substorm Onset in the Magnetotail (A. Nishida) *ELECTROMAGNETIC AND ELECTROSTATIC INSTABILITIES ON AURORAL FIELD LINES* A Review of Electrostatic Wave Measurements on Auroral Magnetic Field Lines (M.C. Kelley) / Diffuse Auroral Precipitation (M. Ashour-Abdalla and C.F. Kennel) / Electromagnetic Plasma Wave Emissions from the Auroral Field Lines (D.A. Gurnett) / Theory of Electromagnetic Waves on Auroral Field Lines (J.E. Maggs) *RAPID AURORAL FLUCTUATIONS AND ASSOCIATED PHENOMENON* Observations of Rapid Auroral Fluctuations (T. Oguti) / Highlights in the Studies of the Relationship of Geomagnetic Field Changes to Auroral Luminosity (W.H. Campbell) / Microburst Precipitation Phenomena (G.K. Parks) *MECHANISMS FOR THE FORMATION OF AURORAL STRUCTURE* Observed Microstructure of Auroral Forms (T.N. Davis) / Birkeland Currents and Auroral Structure (H.R. Anderson) / Relationships between Particle Precipitation and Auroral Forms (J.L. Burch and J.D. Winningham) / Photometric Investigation of Precipitating Particle Dynamics (S.B. Mende) / Generation Mechanisms for Magnetic-Field-Aligned Electric Fields in the Magnetosphere (C.-G. Fälthammar) / Review of Auroral Currents and Auroral Arcs (G. Atkinson) / Acceleration Mechanisms for Auroral Electrons (D.W. Swift) / Subject Index

AEPS Vol. 5

Hard cover edition to Journal of Geomagnetism and Geoelectricity Vol. 30, No. 5, 1978

Proceedings of IAGA/IAMAP Joint Assembly, August 1977, Seattle, Washington

TECTONOMAGNETICS AND LOCAL GEOMAGNETIC FIELD VARIATIONS

Edited by M. Fuller, M.J.S. Johnston, and T. Yukutake

Contents Symposium on Tectonomagnetics and Small Scale Secular Variations Held at the IAGA/IAMAP Joint Assembly at Seattle on Tuesday, August 22nd, 1977 (V.A. Shapiro and M.J.S. Johnston) / Tectonomagnetic Studies in Tajikstan (Yu. P. Skovorodkin, L.S. Bezuglaya, and T.V. Guseva) / An Attempt to Observe a Seismomagnetic Effect during the Gazly 17th May 1976 Earthquake (V.A. Shapiro and K.N. Abdullabekov) / Secular Variation Anomalies and Aseismic Geodynamic in the Urals (V.A. Shapiro, A.L. Aleinkov, A.A. Nulman, V.A. Pyankov, and A.V. Zubkov) / Geomagnetic Investigations in the Seismoactive Regions of Middle Asia (V.A. Shapiro, A.N. Pushkov, K.N. Abdullabekov, E.B. Berdaliev, and M.Yu. Muminov) / Local Magnetic Field Variations and Stress Changes Near a Slip Discontinuity on the San Andreas Fault (M.J.S. Johnston) / Geomagnetic Secular Variation Anomalies in the GDR (W. Mundt) / Noise Reduction Techniques for Use in Determining Local Geomagnetic Field Changes (R.H. Ware and P.L. Bender) / Local Variations in Magnetic Field, Long-Term Changes in Creep Rate, and Local Earthquakes along the San Andreas Fault in Central California (B.E. Smith, M.J.S. Johnston, and R·O. Burford) / Geomagnetic Induction Study of the Seismically Active Fault along the Southwestern Coast of the Sea of Japan (J. Miyakoshi and A. Suzuki) / Time Dependence of Magnetotelluric Fields in a Tectonically Active Region in Eastern Canada (R.D. Kurtz and E.R. Niblett) / Piezomagnetic Response with Depth, Related to Tectonomagnetism as an Earthquake Precursor (R.S. Carmichael) / Magnetic Susceptibility of Magnetite under Hydrostatic Pressure, and Implications for Tectonomagnetism (A.A. Nulman, V.A. Shapiro, S.I. Maksimovskikh, N.A. Ivanov, J. Kim, and R.S. Carmichael) / Effect of Uniaxial Stress upon Remanent

Magnetization: Stress Cycling and Domain State Dependence (J. REVOL, R. DAY, and M. FULLER) / On the Measurement of Stress Sensitivity of NRM Using a Cryogenic Magnetometer (T.L. HENYEY, S.J. PIKE, and D.F. PALMER)

AEPS Vol. 6

Hard cover edition to Journal of Physics of the Earth Vol. 26, Supplement, 1978 (Not included in regular issues)

Proceedings of the International Conference on Geodynamics of the Western Pacific-Indonesian Region, March 1978, Tokyo

GEODYNAMICS OF THE WESTERN PACIFIC

Edited by S. UYEDA, R.W. MURPHY, and K. KOBAYASHI

Contents Plate Tectonic Evolution of North Pacific Rim (W.R. DICKINSON) / Speculations on Mountain Building and the Lost Pacific Continent (A. NUR and Z. BEN-AVRAHAM) / Benioff Zones, Absolute Motion and Interarc Basin (F.T. WU) / Oceanic Crust in the Dynamics of Plate Motion and Back-Arc Spreading (Y. IDA) / Basic Types of Internal Deformation of the Continental Plate at Arc-Arc Junctions (K. SHIMAZAKI, T. KATO, and K. YAMASHINA) / Fault Patterns in Outer Trench Walls and Their Tectonic Significance (G.M. JONES, T.W.C. HILDE, G.F. SHARMAN, and D.C. AGNEW) / Motion of the Pacific Plate and Formation of Marginal Basins: Asymmetric Flow Induction (R.C. BOSTROM) / The Relationship between Volcanic Island Genesis and the Indo-Australian Pacific Plate Margins in the Eastern Outer Islands, Solomon Islands, South-West Pacific (G. Wyn HUGHES) / Upper Mantle Velocity Structure in the New Hebrides Island Arc Region (K.L. KAILA and V.G. KRISHNA) / Upper Mantle Velocity Structure in the Tonga-Kermadec Island Arc Region (K.L. KAILA and V.G. KRISHNA) / Morphology and Structure of the Southern Part of the New Hebrides Island Arc System (J. DANIEL) / Paleomagnetic Evidence for the Rotation of Seram, Indonesia (N.S. HAILE) / A Late Miocene K-Ar Age for the Lavas of Pulau Kelang, Seram, Indonesia (R.D. BECKINSALE and S. NAKAPADUNGRAT) / A Survey of Paleomagnetic Data on Mexico (S. PAL) / Southeast Asian Tin Granitoids of Contrasting Tectonic Setting (C.S. HUTCHISON) / Seismicity, Gravity and Tectonics in the Andaman Sea (R.K. VERMA, M. MUKHOPADHYAY, and N.C. BHUIN) / Focal Mechanisms and Tectonics in the Taiwan-Philippine Region (T. SENO and K. KURITA) / Recent Tectonics of Taiwan (F.T. WU) / Tectonics of the Ryukyu Island Arc (K. KIZAKI) / Explosion Seismic Studies in South Kyushu Especially around the Sakurajima Volcano (K, ONO, K. ITO, I. HASEGAWA, K. ICHIKAWA, S. IIZUKA, T. KAKUTA, and H. SUZUKI) / Two Types of Accretionary Fold Belts in Central Japan (Y. OGAWA and K. HORIUCHI) / Permain and Triassic Sedimentary History of the Honshu Geosyncline in the Tamba Belt, Southwest Japan (D. SHIMIZU, N. IMOTO, and M. MUSASHINO) / Thermal Structure of the Sanbagawa Metamorphic Belt in Central Shikoku (S. BANNO, T. HIGASHINO, M. OTSUKI, T. ITAYA, and T. NAKAJIMA) / Shimanto Geosyncline and Kuroshio Paleoland (T. HARATA, K. HISATOMI, F. KUMON, K. NAKAZAWA, M. TATEISHI, H. SUZUKI, and T. TOKUOKA) / Magnetic Stratigraphy of the Japanese Neogene and the Development of the Island Arcs of Japan (N. NIITSUMA) / Regional Characteristics and Their Geodynamic Implications of Late Quaternary Tectonic Movement Deduced from Deformed Former Shorelines in Japan (Y. OTA and T. YOSHIKAWA) / Magnetic Anomalies and Tectonic Evolution of the Shikoku Inter-Arc Basin (K. KOBAYASHI and M. NAKADA) / A Compilation of Magnetic Data in the Northwestern Pacific and in the North Philippine Sea (N. ISEZAKI and H. MIKI) / Collision of the Izu-Bonin Arc with Central Honshu: Cenozoic Tectonics of the Fossa Magna, Japan (T. MATSUDA) / Flow under the Island Arc of Japan and Lateral Variation of Magma Chemistry of Island Arc Volcanoes (M. Toriumi) / Seismic Activity and Pore Pressures across Island Arcs of Japan (N. Fujii and K. Kurita) / Aseismic Belt along the Frontal Arc and Plate Subduction in Japan (K. Yamashina, K. Shimazaki, and T. Kato) / Tsunamicity of Sanriku Depends on Subduction Tectonics (Wm. M. Adams) / Seismic Studies of the Upper Mantle beneath the Arc-Junction at Hokkaido: Folded Structure of Intermediate-Depth Seismic Zone and Altenuation of Seismic Waves (T. Moriya) / Sedimentary Patterns in Apparent Back-Arc Basins: A Case Study of the Neogene Sequence in Northwestern Hokkaido, Japan (H. Okada) / Velocity Anisotropy in the Sea of Japan as Revealed by Big Explosions (H. Okada, T. Moriya, T. Masuda, T. Hasegawa, S. Asano, K. Kasahara, A. Ikami, H. Aoki, Y. Sasaki, N. Hurukawa, and K. Matsumura) / Geodynamics of the North-Eastern Asia in Mesozoic and Cenozoic Time and the Nature of Volcanic Belts (L.M. Parfenov, I.P. Voinova, B.A. Natal'in, and D.F. Semenov) / The Crustal Structure and Origin of the Basins of Japan Sea and Some Other Seas of the Circum-Pacific Mobile Belt (P.N. Kropotkin) / Major Strike-Slip Faults and Their Bearing on Spreading in the Japan Sea (K. Otsuki and M. Ehiro) / Significant Eruptive Activities Related to Large Interplate Earthquakes in the Northwestern Pacific Margin (M. Kimura) / A Mechanism to Explain the Earthquakes around Japan by the Process of Partial Melting (M. Hayakawa and S. Iizuka) / The Formation of Intermediate and Deep Earthquake Zone in Relation to the Geologic Development of East Asia since Mesozoic (Y. Suzuki, K. Kodama, and T. Mitsunashi) / Subject Index / Geographical Index

AEPS Vol. 7

Hard cover edition to Journal of Geomagnetism and Geoelectricity Vol. 31, No. 3, 1979

Proceedings of IAGA/IAMAP Joint Assembly August 1977, Seattle, Washington

ELECTRIC CURRENT AND ATMOSPHERIC MOTION

Edited by S. Kato and R.G. Roper

Contents Electrodynamics of the Ionosphere from Incoherent Scatter: A Review (M. Blanc)/ Long-Period Waves in Mesospheric Winds at Saskatoon (52°N) (A.D. Belmont and G.D. Nastrom) / Solar Tidal Wind Structures and the *E*-Region Dynamo (J.M. Forbes and H.B. Garrett) / Dynamics of Severe Storms through the Study of Thermospheric-Tropospheric Coupling (R.J. Hung and R.E. Smith) / Coordinated Measurements of *E*-Layer Drifts (E.S. Kazimirovsky and V.D. Kokourov) / On an Origin of Ultra Long Period (Several Days) of Geomagnetic Fluctuations (T. Kitamura) / IMF and Lower Thermospheric Currents and Motions: A Review (S. Matsushita) / Abnormal Features of the Regular Daily Variation S_R (P.N. Mayaud) / Rocket Measurements of Annual Mean Prevailing, Diurnal and Semi-Diurnal Winds in the Lower Thermosphere at Mid-Latitudes (D. Rees) / Mid-Latitude Winds and Electric Fields in the Lower Thermosphere and Their Relationship with the Global S_q Ionospheric Current System (D. Rees) / Ionospheric Wind Dynamo Theory: A Review (A.D. Richmond) / The Quiet-time Equatorial Electrojet and Counter-Electrojet (R.T. Marriott, A.D. Richmond, and S.V. Venkateswaran) / Equatorial Electrojet and S_q Current System —Part I (J. W. MacDougall) / Equatorial Electrojet and S_q Current System—Part II (J.W. MacDougall) / Results from *in situ* Measurements of Ionospheric Currents in the Equatorial Region—I (S. Sampath and T.S.G. Sastry) / Depth of Non-Conducting Layer in the Indian Ocean Region around Thumba, Derived from *in situ* investigations of Equatorial Electrojet—II (S. Sampath and T.S.G. Sastry) / AC Electric Fields Associated with the Plasma Instabilities in the Equatorial Electrojet—III (S. Sampath and T.S.G. Sastry) / Electric Potential

Difference between Conjugate Points in Middle Latitudes Caused by Asymmetric Dynamo in the Ionosphere (N. FUKUSHIMA) / Results of Wind Velocity Measurements at Middle and High Latitudes by the Meteor Radar Method (I.A. LYSENKO, A.D. ORLYANSKY, and Yu. I. PORTNYAGIN) / A Comparison between Radio Meteor and Airglow Winds (G. HERNANDEZ and R.G. ROPER) / Subject Index

AEPS Vol. 8

Hard cover edition to Journal of Physics of the Earth Vol. 27, Supplement, 1979 (Not included in regular issues)

STRUCTURE OF TRANSITION ZONE

Edited by S. ASANO

Contents Crust and Upper Mantle Structure beneath Northeastern Honshu, Japan as Derived from Explosion Seismic Observations (S. ASANO, H. OKADA, T. YOSHII, K. YAMAMOTO, T. HASEGAWA, K. ITO, S. SUZUKI, A. IKAMI, and K. HAMADA) / Regionality of the Upper Mantle around Northeastern Japan as Revealed by Big Explosions at Sea. I. SEIHA-1 Explosion Experiment (H. OKADA, S. ASANO, T. YOSHII, A. IKAMI, S. SUZUKI, T. HASEGAWA, K. YAMAMOTO, K. ITO, and K. HAMADA) / On the Junction Character of the Continental and the Oceanic Lithospheric Blocks in the Kamchatka Region (G.I. ANOSOV, S.K. BIKKENINA, V.I. FEDORCHENKO, A.A. POPOV, K.F. SERGEEV, and V.K. UTNASIN) / New Evidences of the Discontinuous Structure of the Descending Lithosphere as Revealed by *ScSp* Phase (Hm. OKADA)/ A Block Velocity Model of the Focal Zone and Adjacent Mantle in the Kurile-Japan Region (R.Z. TARAKANOV) / Geological Structure of the Southwestern Okhotsk Sea Area (S.L. SOLOVIEV, M.L. KRASNY, O.A. MELNIKOV, Yu.A. PAVLOV, E.I. POPOV, S.S. SNEGOVSKOY, I.K. TUENOV, and B.I. VASILIEV) / Structure and Geological Nature of the Kuril Abyssal Basin in the Okhotsk Sea (I.K. TUEZOV, B.I. VASILIEV, M.L. KRASNY, Yu.A. PAVLOV, and E.I. POPOV) / The Daito Ridge Group and the Kyushu-Palau Ridge—with Special Reference to the Tectonics of the Philippine Sea— (T. SHIKI, Y. MISAWA, and I. KONDA) / Heat Flow in the Hokkaido-Okhotsk Region and Its Tectonic Implications (S. EHARA) / Geothermal Investigations Carried out in the Northwestern Sector of the Pacific Mobile Belt (O.V. VESELOV, N.A. VOLKOVA, V.V. SOINOV, and Y.D. EREMIN) / Deep Electric Conductivity Study in the Asia-to-Pacific Transition Zone (L.L. VANYAN, V.V. BORETZ, A.M. LYAPISHEV, B.E. MARDERFELD, A.V. RODIONOV, and V.N. VERKHOVSKY) / Electrical Conductivity Structure beneath the Japan Island Arc by Geomagnetic Induction Study (J. MIYAKOSHI) / Geomagnetic Anomalies of the Japan Sea by Japanese and Russian Magnetic Data (N. ISEZAKI, M. YASUI, and S. UYEDA) / Recent Crustal Movements in Primorie and Sakhalin (V.K. ZAKHAROV, B.A. BELOUSOV, V.P. SEMAKIN, N.F. VASILENKO, and G.G. YAKUSHKO) / Subject Index